Y0-BYA-396

# WEATHER
## *of*
# ALBERTA

## Bill Hume

LONE
PINE

Lone Pine Publishing

© 2008 by Lone Pine Publishing
First printed in 2008  10 9 8 7 6 5 4 3 2 1
Printed in China.

All rights reserved. No part of this work covered by the copyrights hereon may be reproduced or used in any form or by any means—graphic, electronic or mechanical—without the prior written permission of the publisher, except for reviewers, who may quote brief passages. Any request for photocopying, recording, taping or storage on information retrieval systems of any part of this work shall be directed in writing to the publisher.

**The Distributor: Lone Pine Publishing**

10145–81 Avenue
Edmonton, AB
T6E 1W9

Website: www.lonepinepublishing.com

**Library and Archives Canada Cataloguing in Publication**

Hume, Bill, 1949-
    Weather of Alberta / Bill Hume.

Includes bibliographical references and index.

ISBN 978-1-55105-602-9

    1. Alberta--Climate.  I. Title.

QC985.5.A53H84 2008     551.657123     C2008-902642-X

*Editorial Director:* Nancy Foulds
*Project Editor:* Gary Whyte
*Editors:* Wendy Pirk, Sheila Quinlan
*Production Manager:* Gene Longson
*Book Design & Production:* Michael Cooke
*Layout:* Michael Cooke, Trina Koscielnuk
*Maps & Charts:* Volker Bodegom
*Illustrations:* Megan Fischer, Gerry Dotto, Trina Koscielnuk, Michael Cooke
*Photo Coordination:* Randy Kennedy
*Cover Design:* Gerry Dotto

Front cover photograph by: JupiterImages Corporation

Photo Credits: Every effort has been made to accurately credit copyright owners. Any errors or omissions should be directed to the publisher for changes in future editions. The photographs in this book are reproduced with the kind permission of the copyright owners. All photo credits are located on p. 239.

PC: *P16*

# Table of Contents

# Dedication

For my wife Judith, who has always been a source of strength and encouragement in the projects we have undertaken.

And to the memory of my father, whose interest in weather inspired his son's career.

# Acknowledgements

Many thanks to those who helped make this book a reality. Much appreciation goes to many at Environment Canada, including Brian Proctor, Dan Kulak, Kitty Wilkes, Ed Hudson, Bob Kochtubjada, Jeff Sowiak, Ron Goodson, David Phillips and Gloria Turnbull, for the productive discussions we had. Jim Renick, Terry Krauss and Geoff Strong were a wealth of knowledge on severe summer storms and weather modification in Alberta. Professors Keith Hage and Ed Lozowsky from the University of Alberta helped ground my early studies in atmospheric science and also provided information that found its way into this book. Markus Kellerhals, Frank Letchford, Kitty Wilkes, Geoff Strong, Jim Renick and Bob Kochtubjada provided helpful reviews of portions of the text.

Shane Kennedy at Lone Pine Publishing provided the leadership and inspiration, Nancy Foulds the direction and Wendy Pirk helped guide the writing. The efforts of Gerry Dotto, Gary Whyte and Randy Kennedy at Lone Pine were much appreciated.

# Introduction

Weather in Alberta covers the whole gamut, from sudden blizzards in August and having to pull out the snow shovel every second day throughout winter, to monsoon-like rains in June and tornadoes in July. In some years, Alberta gets the full range of bizarre weather, whereas other years are totally unmemorable for most people in the province. The weather is a frequent topic of conversation for Albertans; we love to discuss how the weather can change from one year to the next and even from day to day. As many Albertans will attest, "if you don't like the weather, just wait 15 minutes." Chinooks have been known to raise temperatures by over 30° C in 30 minutes, accompanied by winds that are nearly calm and then suddenly gust to over 100 kilometres per hour.

The weather has long been part of Alberta's heritage. In the written history of settlement in the province, the most memorable accounts are the stories of hardship inflicted by Mother Nature. There are many tales of winter blizzards, of rainstorms that flooded towns and cities alike and of dust storms that moved Alberta's topsoil into neighbouring Saskatchewan. But these are not just the stories in history books. Weather history is still being written in the province. The relatively recent Edmonton and Pine Lake tornadoes were tragic for many and left an indelible imprint in the minds of thousands of others.

## Ecoregions of Canada

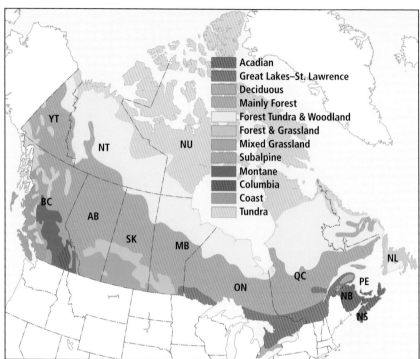

Acadian
Great Lakes–St. Lawrence
Deciduous
Mainly Forest
Forest Tundra & Woodland
Forest & Grassland
Mixed Grassland
Subalpine
Montane
Columbia
Coast
Tundra

Alberta's weather is moulded by two key factors. First, a belt of mid-latitude westerly winds blows across the province, and weather systems caught up in the flow pattern make Alberta a frequent battleground. Second, the Rocky Mountains exert a major control on the character and behaviour of weather systems. Much of the moisture is wrung out of the air masses coming in from the west, and the mountains provide a barrier that helps to funnel cold air over the province from the northwest and to stimulate heavy precipitation. The long-term effect of atmospheric behaviour gives rise to a climatic range that has yielded a range of ecosystems across the province. The Ecoregions of Alberta map on p. 7 shows this distribution of ecoregions.

Grasslands of the southeast give way to a relatively narrow aspen parkland region through central Alberta. The northern half of the province is boreal forest. The Cordilleran Region extends through a narrow band along the southwest border.

Despite Alberta's wild weather variation, no Canadian weather records have been set in the province. Based on a compilation of weather data for major centres across Canada, the province lived up to its reputation as sunny Alberta: it was the sunniest. However, in most of the 40 weather categories evaluated, the eight Alberta centres that were examined placed in the middle of the pack. Overall, Alberta's climate is ranked as the most comfortable in Canada.

# Ecoregions of Alberta

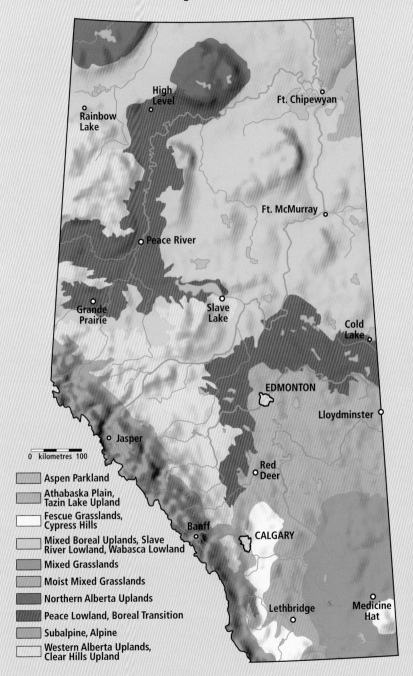

0  kilometres  100

- Aspen Parkland
- Athabaska Plain, Tazin Lake Upland
- Fescue Grasslands, Cypress Hills
- Mixed Boreal Uplands, Slave River Lowland, Wabasca Lowland
- Mixed Grasslands
- Moist Mixed Grasslands
- Northern Alberta Uplands
- Peace Lowland, Boreal Transition
- Subalpine, Alpine
- Western Alberta Uplands, Clear Hills Upland

Rainbow Lake

High Level

Ft. Chipewyan

Peace River

Ft. McMurray

Grande Prairie

Slave Lake

Cold Lake

EDMONTON

Lloydminster

Jasper

Red Deer

Banff

CALGARY

Lethbridge

Medicine Hat

Meteorology is the science that defines how the atmosphere behaves. This science is based on physics and mathematics, and it immediately becomes complex, if not downright confusing. This book attempts to explain the operation of the atmosphere with narrative descriptions and diagrams, but it does not include any equations. It is said that a picture is worth a thousand words, so the images included in these pages should sufficiently illustrate meteorology's many scientific complexities. Those interested in a more rigorous treatment of the science are better off referring to any number of scientific texts, some of which are included in a list of references at the end of the book. There are also many excellent web sites that can be investigated.

The Environment Canada site (www.msc.ec.gc.ca/weather) provides much information on atmospheric science. Weather forecasts are available at the Canadian Weather Network site (www.theweathernetwork.com) and at the Environment Canada site (www.weatheroffice.gc.ca).

This book will provide you with an understanding of weather and many of the associated phenomena, with an emphasis on Alberta and its varied climate. Within these pages, we discuss the atmosphere—its structure and how it works—as well as other atmospheric phenomena.

Alberta may experience less cloud cover overall than the other provinces,

but we still see our fair share, so the cloud types that we encounter in our province, the precipitation process and the most significant storm types to which we are subjected are covered. And no discussion of storms would be complete without a mention of atmospheric electricity, including the high atmospheric aurora borealis often viewed across the province.

Every Albertan is familiar with wind, from the bone-chilling arctic winds of winter, felt throughout the province, to the howling Chinook winds coming down from the Rockies in the southwest. Read on to learn about the mechanics of wind, as well as some of the different types of winds that can be expected in Alberta.

Because weather has such a huge impact on our day-to-day lives, people throughout time have been keeping an eye on the skies. Weather observation is a science that has a long history; records of formal weather observation in Canada date back to the 18th century, and some of the earliest formal measurements in Alberta were taken in 1843 at Fort Chipewyan and Fort Edmonton. In this book, descriptions of the measurement systems and equipment used to observe weather in the province are provided.

However, simply observing the weather isn't enough for most of us; we want to know what conditions we can expect in the upcoming days, and at times, we've even tried to influence weather conditions to better suit our needs. Both weather forecasting and weather modification science and its practice in Alberta are discussed in later chapters.

Venturing beyond the day-to-day weather phenomena, the final chapters examine two broad atmospheric issues: air quality and how the atmosphere deals with air pollution; and the long-term future of our atmosphere, including changes that we have induced on the protective ozone layer and the ever contentious issue of climate change.

I hope that you find this an informative explanation of weather phenomena in Alberta.

# Chapter One: The Atmosphere

> **Some are weather-wise,
> some are otherwise.**
> —Benjamin Franklin

Compared to the dimension of the earth, the atmosphere is just a thin skin of gas. But it is complex; it contains a number of constituents, has a complicated structure and has many protective properties without which life as we know it could not survive. As we note in pictures taken from space, the blue atmosphere has considerable associated structure, which is visible because of the presence of clouds. When observed with time-lapse photography, movement and changes in the cloud patterns suggest that there is indeed some structure. There may be large areas with organized shape. White cloud masses swirl, expand and shrink, while cloud bands in the shape of lines and arcs move across the earth's surface.

Although the motion may appear chaotic, it actually closely follows paths and patterns that are described by rules governing the complex science of fluid dynamics. These are the same laws that describe the motion of fluids flowing in channels and pipes, or the motion of air around an aircraft wing. These motions all have a chaotic nature to them—they have some degree of associated turbulence. But the flow mostly behaves in an explainable manner, and its motion can be predicted.

# The Composition of the Atmosphere

The composition of the atmosphere is 21 percent oxygen and 78 percent nitrogen. The remaining one percent includes a wide variety of other gases, including water vapour, carbon dioxide, the inert gas argon and a host of others, some naturally occurring and some human made. Oxygen and nitrogen are well mixed throughout the atmosphere. Many of the other gases are less uniformly distributed, a result of their tendency to chemically react with other gases, to change phase, or to remain in the region of the atmosphere near their emission source. Gravity keeps the gaseous envelope to within a depth of a few hundred kilometres of the earth's surface. The interaction between molecules pushes them apart, preventing the gas from collapsing into a lump of oxygen and nitrogen at the surface. This interaction is known as internal pressure. The molecules constantly collide with each other and with other objects. Near sea level, at temperatures near 20° C, the individual molecules move at about 500 metres per second. Molecules in the lower atmosphere are dense—near the earth's surface, 1 cubic centimetre contains about 25 sextillion molecules (or 25 followed by 21 zeroes!).

To picture an analogy for pressure, imagine yourself throwing balls at a bathroom scale. Each time a ball hits the scale, the scale registers a slight reading, which depends on the speed and weight of the ball that was thrown. When many balls are thrown over a fixed period of time, say one second, the weight indicator on the scale does not have time to return to zero—a steady reading, or pressure of the colliding balls, is displayed. It is the same with the atmosphere. Over a 1-square-centimetre surface, there are hundreds of millions of molecules, each

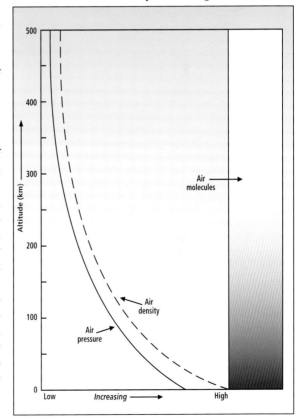

## Variation of Atmospheric Pressure and Density with Height

Fig. 1-1

11

with a tiny mass, striking per second. This molecular pressure pushing in all directions, including upward, counteracts the downward pull of gravity. The mass of all the molecules in a vertical column with a surface area of 2.5 by 2.5 centimetres extending from sea level to the top of the atmosphere is about 6.7 kilograms. This internal pressure, measured as a force per unit area, pushes back the earth's gravitational attraction to the molecules. In atmospheric science, the pressure force is called the millibar (mb).

The higher one goes in the atmosphere, the fewer molecules there are, and the weight of the molecules, as well as the corresponding internal pressure, is reduced. There is a mathematical equation that defines how atmospheric pressure decreases with height. This equation shows (and measurement confirms) that at a height of 5 kilometres, atmospheric pressure is reduced to about 50 percent of its value at sea level, and at a height of 10 kilometres, to about 30 percent.

## Seasons as a Result of Earth's Orbit

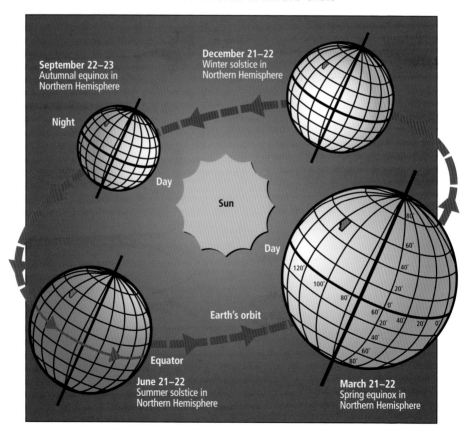

Fig. I-2 The tilt of the earth's axis of rotation gives us seasonal changes.

## Solar Energy Reflected by Earth's Surfaces

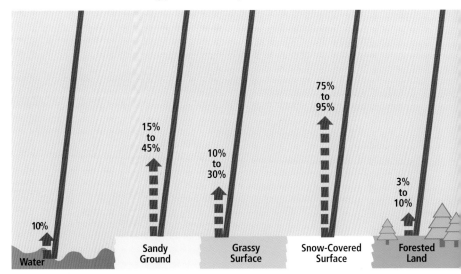

*Fig. I-3* Different surfaces reflect varying amounts of the sun's radiation.

## Solar Input and its Variation

The sun is the primary driver of weather on earth, but it is the tilt of the earth's axis of rotation that gives us the seasonal changes. This angle of inclination (23 degrees) remains constant as the earth proceeds on its annual trek around the sun. The orbit is not quite circular—it is elliptical. When the earth's axis is tilted away from the sun, and Alberta is in the middle of winter, the earth is at its closest point to the sun.

The sun's radiation covers the full electromagnetic spectrum, from the very short x-rays, through the visible light rays to the much longer radio wave frequencies. However, the amount of radiation differs across the spectrum. About 40 percent of the sun's energy is emitted in the visible and near infrared frequencies. The amount of solar radiation that reaches the earth's surface is less than what is incident at the top of the atmosphere. Ozone molecules in the high atmosphere absorb most of the ultraviolet energy, whereas much of the incoming infrared energy is absorbed in the lower levels of the atmosphere.

Clouds and snow surfaces can reflect up to 95 percent of incident radiation, whereas water surfaces reflect about 10 percent. Sand, grass surfaces and forests reflect between 10 and 30 percent. These reflections depend on the angle of incidence of the radiation—the lower the angle of incidence, the greater the amount of reflection. Overall, 43 percent of the solar radiation entering the top of the atmosphere is reflected and scattered back to space. Some absorption by dust and gases, including water vapour, further depletes the incoming radiation, with the result that about 40 percent is absorbed by the

13

## Energy Transfers in the Earth-Atmosphere System

Sun

Absorbed by dust, gases and clouds

Scattered by atmosphere into space

Radiated into space from the atmosphere

Direct, diffuse and scattered

Reflected by earth's surface

Reflected from clouds into space

Radiated into space from earth's surface

Surface

Absorbed by earth's surface

Absorbed energy converts to heat, which warms the air, evaporates water and melts snow and ice

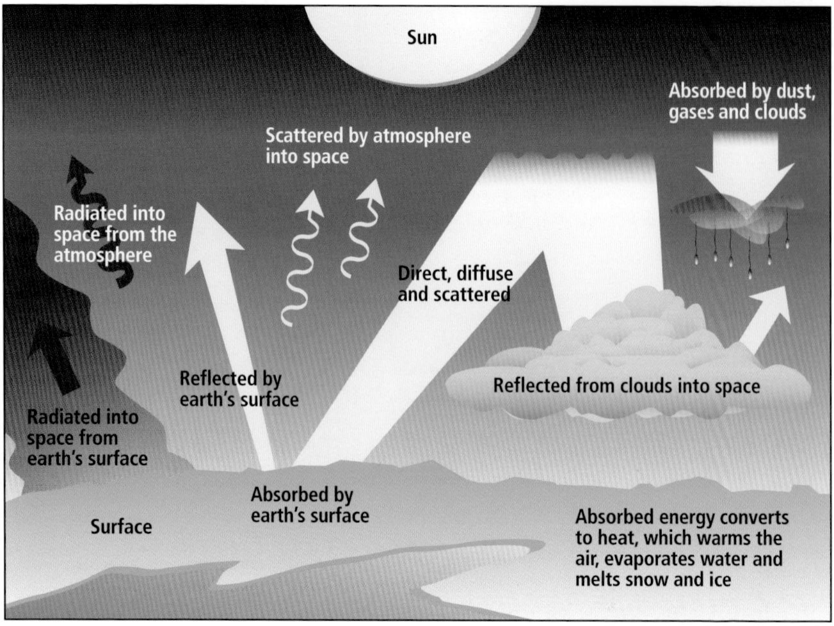

*Fig. 1-4* Some solar energy is reflected and scattered back into space. The remainder is absorbed in the earth-atmosphere system, which radiates energy. Overall, it is almost in balance.

earth's surface. All bodies that absorb energy are heated, and these warm bodies in turn emit radiation at the infrared frequency. This energy goes back out into space. There is a balancing act in constant play between the incoming and outgoing energy, and the average temperature of the earth is reasonably steady over time. Indications of an imbalance in this energy account give rise to the global warming concept, which is discussed in a later section of this book.

The Alberta land surface, of course, plays a role in the global energy balance. Alberta's land surface is a mixture of forest cover and grasslands with a relatively small proportion of water surfaces and non-vegetated terrain in the mountains and badlands. The rates of absorption of solar energy vary from the north to the south and, most significantly, with the seasons. The mountains in western Alberta play a large role in drying out moist air moving from the west. The mountains also act to help "dam" the movement of cold, dense air toward the west. As a result, Alberta is subject to large temperature extremes, from the highs under cloudless summer skies to the frigid cold snaps that occur with low sun angles and clear nights in winter.

## Vertical Variation of the Atmosphere

As already noted, atmospheric pressure decreases with altitude, but what about other measures? Temperature is the most significant parameter. The behaviour of temperature is used to define the layers of the atmosphere. Rates of temperature change with height are known as thermal lapse rates.

At Alberta's latitude, temperatures in the lower atmosphere generally decrease from their highest values near the earth's surface to their lowest values at altitudes of about 10 to 12 kilometres. This layer of the atmosphere, characterized by decreasing temperature, or negative lapse rate, is called the troposphere, and it is here that all active weather occurs. The warmth at the bottom of the troposphere is a result of the warmth of the land and water surface combined with the heat contained in water vapour. The temperature decrease stops abruptly or pauses at a level known as the tropopause, where temperatures are around -50° to -60° C.

Moving upward in the atmosphere, the temperature slowly increases through the layer known as the stratosphere, until it reaches its maximum at a height of about 50 kilometres. The stratosphere contains most of the atmosphere's ozone, and warming through this layer is mostly a result of the absorption of the sun's ultraviolet radiation. The maximum temperature is often in excess of 0° C at this height. This maximum temperature level is known as the stratopause. Thereafter, temperatures decrease again through what is known as the mesosphere, approaching a minimum value near -90° C at the mesopause, which lies at a height of approximately 80 to 90 kilometres and has an atmospheric pressure of about 0.01 mb. Above the mesopause, temperatures increase

## Vertical Structure of the Atmosphere

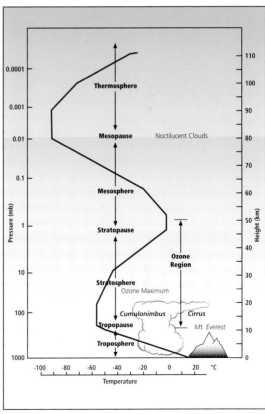

Fig. I-5 Temperature changes with height and defines the different layers of the atmosphere. All the active weather is in the troposphere.

with height through a layer known as the thermosphere. In spite of low density within the thermosphere, the tenuous atmosphere can still affect drag on spacecraft at altitudes of 250 kilometres.

Moisture in the atmosphere plays a major role in the formation and development of weather. Virtually all the moisture is contained in the troposphere and is in its highest concentrations at the bottom of this atmospheric layer. The heating of this layer by the sun results in turbulent mixing, which distributes moisture and heat to higher levels. Moisture is also transported upward in the troposphere by the more gradual processes associated with what is known as frontal lift. Only rarely are the turbulent mixing or the frontal lift processes vigorous enough to drive moisture through the tropopause.

*Convection starts in the lowest layer of the atmosphere, where warm air rises. As it moves higher, the air expands and cools. Nearby air sinks back toward the ground.*

## Atmospheric Circulation

Horizontal variation in atmospheric parameters has much more complexity than vertical variation. The earth's atmosphere can be thought of as a heat engine that is driven by the sun. In its simplest form, the basic function of the heat engine is to move heat from the areas where it is most intense—in the equatorial zone—toward the cooler regions near the north and south poles.

When scientists started to think about atmospheric circulation, they developed simple models to explain apparently organized phenomena. George Hadley developed a simple model in 1735 to explain the elements of the easterly trade winds that are observed in the tropics. Hadley noted that the equatorward motion of air in the easterly trade winds must be balanced in some fashion to prevent the accumulation of mass in the equatorial zone. He postulated that upward motion and then poleward motion of air would result. However, the simple model he proposed does not fully explain the circulation and associated weather patterns observed around the globe.

A model of atmospheric circulation developed by Rossby in 1941 helps explain some of the complexity in the mechanism whereby heat is transported away from the equator toward the poles. This model, as depicted in Figure 1-6, has a considerable element of reality, especially for Albertans at so-called mid-latitudes. In this model, the Hadley circulation shows flow toward the equator, ascent and return flow aloft, and then subsidence in the subtropics. This is a direct cell, driven by heat. The subtropics, where air is descending, are regions of generally high pressure (or anticyclonic circulation) and sunny skies, which are well known to the "snow birds" who flee to those latitudes (from 20° to 30° N of the equator) for winter vacations. Another direct circulation cell is present at high latitudes (60° to 90° N). Here, cold air moves away from the pole toward the south. There is a general area of subsidence near the pole (surface anticyclone or high pressure), and a compensating northward flow aloft. Between

## Hadley Circulation

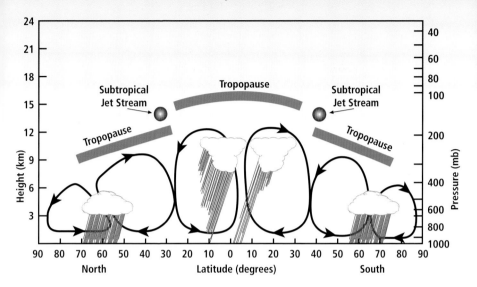

Fig. 1-6 A cross-section view of the atmosphere shows a three-cell circulation on each side of the equator. There is active weather in regions of general ascent.

the two direct cells is an indirect cell in which surface air is transported northward and air aloft southward. Warm air from the south and cold air from the north meet at mid-latitudes. This is the zone where mid-latitude low pressure systems persist. This zone of air mass mixing, convergence and uplift associated with the mid-latitude low pressure systems provides all the key ingredients for production of the weather phenomena that Albertans observe.

The simple circulation processes described do not take into account the rotation of the earth on its axis. This rotation has two important consequences. First, the zone of maximum heating moves from east to west as the earth rotates on its axis. Second, the earth's rotation also causes an apparent deflection of the airflow. On a rotating surface,

an apparent force deflects any moving body. Gustave-Gasparde Coriolis first described this effect mathematically in 1835. North of the equator, the deflection is to the right, and south of the equator the deflection is to the left. The magnitude of the apparent force is proportional to the speed that the body moves.

Air parcels move because of the force of pressure. Two bodies of air that are

**Blow winds and crack your cheeks! Rage, blow,**

**You cataracts and hurricanes, spout**

**Till you have drenched our steeples, drowned the cocks!**

—Shakespeare, *King Lear*

separated horizontally with different temperatures have different densities. Essentially, the colder, higher density air mass has a higher pressure than the adjacent warm, less dense air. Pressure differences between air masses causes air parcels between the masses to move toward lower pressure. This is known as the horizontal pressure gradient force. And the greater the pressure difference (the larger the pressure gradient), the greater the force and therefore the faster the air parcels move. At the same time, the Coriolis force causes a deflection of the air parcels. If we ignore other forces, such as friction or the centrifugal acceleration in curved flow, the Coriolis force pretty well balances the pressure gradient force, and the air parcels move along lines of constant pressure.

## Representation of Weather Systems

It turns out that the most convenient way to depict weather systems is by means of pressure patterns. Moving along a constant height surface, there is a reasonably large variation of pressure in the horizontal. For example, on a map drawn at sea level, pressure can vary by more than 100 mb. Atmospheric pressure is measured at many locations around the world. When pressures measured at the same time are plotted on a map, areas of high or low pressure can be discerned. An "L" marks the centre of an area with pressure lower than its surroundings, whereas an "H" marks the centre of an area of relatively high pressure. Lines joining locations that have the same pressure are called isobars. The greater the difference between the high and low pressure areas, the greater the number of isobars between those areas, and the greater the wind speed. So basically, a weather map is a depiction of pressure patterns with high and low pressure areas.

A number of other parameters are measured at the same time as pressure, and these can also be plotted on the weather map. When the patterns of all these parameters are inspected, many relationships become apparent. Low pressure areas are generally associated with areas of cloud, and often with precipitation.

## Wind Deflection as a Result of Coriolis Effect

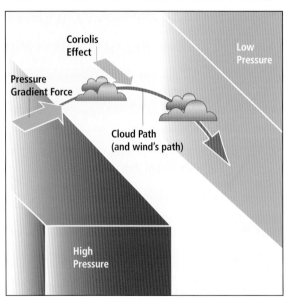

*Fig. 1-7* Air moving from high pressure to low pressure is deflected to the right in Northern Hemisphere.

## Linkages Between Upper Air and Surface Patterns

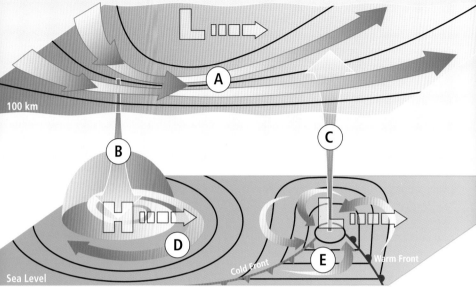

Fig. 1-8

A - wind flows around a moving low centre in the upper atmosphere
B - air subsides from an area of converging air aloft toward the ground
C - air ascends from a surface low toward an area of diverging air aloft
D - air spirals outward and clockwise around a surface high
E - air spirals inward and counterclockwise around a surface low

High pressure areas are associated with relatively cloud-free areas. Relatively warm temperatures are usually near low pressure centres, whereas relatively cool temperatures are in areas of high pressure.

Maps of conditions at other levels in the atmosphere can also be drawn. When depicting weather patterns at upper levels, the relationship between pressure and height is used. Rather than depicting pressure patterns at a constant height, constant pressure levels are chosen and the height of that pressure level above sea level is depicted. Meteorologists examine pressure patterns at several

*A synoptic chart is a map that displays the locations of highs, lows and fronts in a particular region.*

pressure levels within the troposphere. Again, patterns in this height field can be analyzed. As one progresses upward in the atmosphere, the patterns become less "busy" than those at the surface. There are usually fewer centres of high or low height than centres of high or low pressure on a surface weather map covering the same area. As well, the patterns

19

of height appear to have a rather uniform shape, often with broad areas that have an undulating or wavelike appearance. The relationships between these patterns, their shapes, and the juxtaposition of patterns at different levels vertically then give the meteorologist a representation of the vertical structure of the atmosphere. This is the essence of synoptic meteorology—understanding the linkages between weather patterns through the depth of the troposphere.

The study of synoptic meteorology has been underway for about 90 years. When surface observations were combined with observations aloft and the patterns were analyzed, it became possible to develop a three-dimensional model of how the atmosphere was structured. The knowledge that the atmosphere has structure and behaves in an understandable way led to a number of theories about how weather patterns moved and changed with time. The science is complex and can be represented by the mathematics of fluid dynamics, which we will not attempt to reiterate—textbooks abound. But we can try to understand the elements of these atmospheric dynamics well enough to understand the formation, movement and prediction of weather systems over Alberta.

Two key concepts play a crucial role in understanding the behaviour of weather in Alberta. The first concept is that of air masses and the so-called frontal regions that separate air masses. The second concept is the link between the motion of weather systems and the atmospheric flow patterns aloft.

# Air Masses and Fronts

Air that remains over a land or ocean surface for some period of time takes on characteristics dependent on the underlying surface. In meteorology, the term air mass is used to describe a large pool of air that has a uniform character. If the land surface is intensely heated by the sun, the overriding air will be warm, and if the underlying surface is snow or ice covered, the air will be cold. If the air is over an ocean, moisture will evaporate into the lower portion of the air mass. Air that remains over central regions of North America during summer is called a continental polar air mass, and Albertans occasionally experience this air mass. When the circulation is more active, air masses stream into Alberta from other regions. Air that originates from the northern Pacific Ocean, known as maritime arctic air, often crosses Alberta. Air from the arctic regions, which is dry and usually cold, is known as a continental arctic air mass. In winter, this air mass gives Albertans the coldest (occasionally record-breaking cold) conditions. Maritime tropical air masses, originating over low latitude regions of the Atlantic or Pacific Ocean, rarely come as far north as Alberta.

*The jet stream was not discovered until World War II, when pilots noticed they lost ground speed when flying against it.*

# Fronts

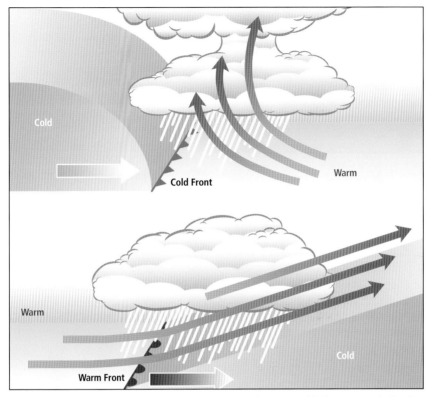

*Fig. 1-9* With a cold front, warm air is displaced by advancing cold, dense air and clouds are formed. With a warm front, warm air overruns the retreating cold air, forming clouds and precipitation.

The boundaries between adjacent air masses are known as fronts. Between WWI and WWII, Norwegian meteorologists established the frontal concept, as it is known, to give some structure to weather systems that influence mid- and high latitudes. This concept is a key building block of modern meteorology. Weather conditions are usually described in relation to the presence, motion and modification of frontal zones. When air from a warm air mass moves against a cooler air mass, the boundary is known as a warm front. When the colder air mass prevails and pushes the warmer air, we have a cold front. As at other locations in the mid-latitudes, Alberta's weather is frequently characterized by the approach and passage of warm and cold fronts.

Another important aspect of the frontal model of the atmosphere is that regions of low pressure frequently form along the frontal boundaries between air masses. The characteristics, behaviour and motion of these low pressure systems are the key to describing present and future weather conditions.

21

## Motion of Weather Systems

The development and movement of frontal storms and weather systems is tied to weather patterns in the mid- to upper troposphere. The patterns differ between the northern and southern hemispheres, and there is little apparent interaction between patterns across the equator. As shown in Figure 1-10, the wind flow at these levels usually displays a wavy, undulating pattern. Meteorologists characterize these patterns as either long waves or short waves.

*Lows bring strong winds, clouds and rain. Highs bring clear skies, perhaps with some high cloud, and can result in very cold weather in winter.*

Long waves are known as planetary waves because they are discernable over the whole planet. The planetary wave, with its characteristic trough and ridge pattern, is continuous around the hemisphere. Over a period of days to weeks, there may be an apparently stable configuration in the planetary pattern with a distance of several thousand kilometres between successive troughs. These planetary waves usually move slowly eastward at a speed of about 10 to 15 kilometres per hour. Planetary waves will, on many occasions, appear to remain stationary, and portions of the pattern may even move to the west, a rare phenomenon known as retrogression.

Short waves, on the other hand, move relatively quickly. They appear as ripples in the planetary wave pattern and move along the long wave pattern from west to east. The speed of these waves can be fairly high, often 40 kilometres per hour or more. Short waves are closely tied to the so-called synoptic scale storms that are 300 to 500 kilometres across. In summary, short waves, with their attendant storm systems, move along, or are steered by, the planetary wave pattern.

The speed of the wind itself within the steering flow is usually much greater than the rate of progression of the short and planetary waves. As well, the amplitude of the short wave ripples usually decreases the higher we go in the atmosphere. The short waves are most apparent at the middle level of the troposphere; therefore meteorologists pay a lot of attention to their shape, rate of motion and other characteristics. The long wave pattern is most evident at the top of the troposphere, and winds are generally strongest at this altitude. These winds are known as the jet stream. Meteorologists often designate the jet stream as the steering flow for short waves and storms.

The location of the ridges and troughs in the planetary wave pattern also plays a role in determining average weather conditions. When the pattern is relatively stable, the associated weather conditions are also stable. Under a ridge, an area of high pressure dominates at the surface. High pressure areas are associated with subsiding, or a downdrift, of air. There is usually little cloud, and fair skies with sunny conditions are prevalent. On the other hand, a trough is associated with unsettled weather. When short waves round the bottom of the planetary trough, the associated weather systems are quite active. A considerable amount

## Wave Pattern in Wind Flow

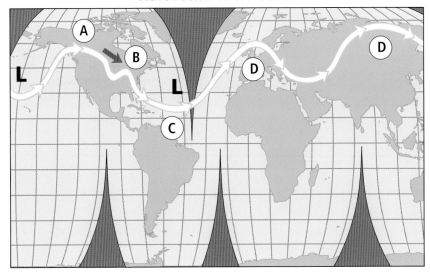

*Fig. 1-10*
A - jet stream
B - short wave moving through ridge
C - trough in planetary wave
D - ridge in planetary wave

of vertical motion may be generated through the mid-troposphere, usually producing cloud, stormy conditions and precipitation. The distance between the trough and ridge in the planetary wave pattern is often 2000 to 3000 kilometres. When the pattern is stable over Canada with the ridge positioned over the west, the trough is positioned over the east. As a result, warm dry conditions will prevail over Alberta, whereas unsettled and perhaps cooler conditions will be prevalent in the east. Of course, the reverse situation also occurs.

## Weather Influences in Alberta

Alberta is located within the mid-latitude zones where air masses and associated fronts are vying for supremacy. The steering flow and associated jet stream are often in close proximity, so Albertans are subjected to the associated weather. But local conditions also play a large role. Perhaps most significant is the presence of mountainous terrain to the west. Maritime polar air directed toward Alberta must go over the Rocky Mountain range. As the air moves to the east, significant lifting of the air mass occurs. Air parcels associated with the maritime polar air mass that originate near sea level off the west coast may be forced to lift several thousand metres. This lifting causes the parcels to cool, usually leading to condensation of the moisture, which is then precipitated out of the air on the upslope ranges. Once the air parcels reach the peak of the mountain range, the desiccated air parcels move down the lee slopes, warming as they move to lower elevations. The rate of warming for the dry air is greater than the rate of cooling for moist, ascending air. As a result, the lifting process over

23

## How an Alberta Clipper is Formed

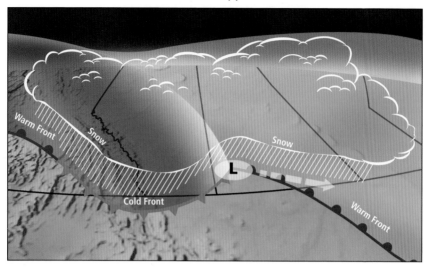

*Fig. 1-11* The formation and movement of an Alberta Clipper

the mountains actually causes a warming. The process just described is the essence of the Chinook phenomenon that occurs all along the lee of the Rockies. Chinooks generally form when the wind flow at mid-tropospheric levels is perpendicular to the Rocky Mountain range. This wind flow is often relatively mild and, at low levels, relatively moist. The onset of Chinook conditions can be quite rapid, and the effect is most pronounced when a cold arctic air mass has been resident for some period of time. The downslope flow starts as the warming progresses to lower elevations along the eastern slopes, "scouring out" the cold air.

Another mountain influence affects the formation of low pressure systems in the lee of the Rocky Mountains. As a slab of air that extends from sea level to the tropopause approaches the west coast and begins to traverse the mountains,

the depth of the air mass decreases—the bottom is pushed up. Winds curving around the base of a low pressure system decelerate on the upward slope. Effectively, a local ridge of high pressure is observed over the mountain range. As the air mass travels to the east of the Rockies, it stretches as it moves down the descending land surface. Low pressure systems may intensify as they move into central Alberta, or a new surface low pressure system may form. This is the lee cyclogenesis phenomenon, which happens whenever the westerly winds are forced to flow over a mountain range. In winter, if there is a frontal zone lying through Alberta, the low pressure system forms along this front. It moves rapidly off to the east and may further develop into a significant winter storm. In eastern North America, these storms are often called "Alberta Clippers." They move rapidly to the east and may have strong winds and snow followed by cold snaps.

# Chapter Two: Clouds

**The cloud never comes from the quarter of the horizon from which we watch for it.**
—Elizabeth Gaskell

For many people, clouds are the most interesting aspect of weather. Our perception of what each day will be like is based on our first glimpse of the sky and the presence or absence of clouds.

Clouds consist of tiny water droplets or ice crystals that occur as a result of the condensation of atmospheric water vapour. Condensation occurs with great difficulty in clean air, so atmospheric water vapour must have a surface upon which it can condense. The formation of cloud particles is therefore dependent on the availability of tiny particles known as condensation nuclei. These condensation nuclei are almost always abundant in the atmosphere, and they arise from dust, pollution particulates, pollen and spores that are stirred into the atmosphere as a result of wind motion and turbulence. In most regions of the lower troposphere, the density is at least 10,000 per cubic centimetre. In industrial areas, concentrations can increase to more than one million nuclei per cubic centimetre. There is no shortage of condensation nuclei available for condensation!

There is an upper limit on how much water vapour can be present in the free atmosphere. This amount depends on the temperature and pressure of the air containing the moisture. Relative humidity is the main way we express water content in the atmosphere. The relative humidity is a ratio of the amount of water vapour in the air compared to the maximum amount of water vapour that the air parcel can contain at the same pressure and temperature. This relative humidity is expressed as a percentage— air that is saturated with water vapour has a relative humidity of 100 percent. So, when clouds are formed, the relative humidity of the air containing the cloud is 100 percent. Another term that is often used to measure atmospheric moisture is the dew point temperature. When unsaturated air is cooled at a constant pressure and without the addition of moisture, this is the temperature at which the air becomes saturated and dew starts to form. The dew point temperature is often a value that is representative of the widespread characteristics of an air mass.

*Cumulonimbus clouds can grow up to 18 kilometres high and can hold more than a half-million tonnes of water.*

For cloud to form, there must be a mechanism for the air to increase its relative humidity to 100 percent. When air that contains water vapour cools, the vapour transforms into cloud droplets. Occasionally, a moist air mass can cool when it moves over a colder surface—this is one of the ways fog forms. Generally,

however, air cools and expands as it ascends, moving into areas of lower pressure. The rate of cooling of the parcel in the lower half of the atmosphere is approximately 10° C per kilometre of ascent. Usually the water vapour remains within the parcel of air and cools at the same rate. When the water vapour cools to a temperature at which condensation starts occurring on condensation nuclei, the air parcel is described as saturated. Its relative humidity is 100 percent, and cloud forms. This altitude is then the base of the cloud.

If the parcel continues to rise, the rate of cooling changes. In fact, the condensing water vapour releases latent heat that was stored during the formation of water vapour by evaporation. This release of latent heat effectively cuts the rate of parcel cooling by about half, which has important consequences for development of convective clouds and thunderstorms.

Ascent of air can occur in a number of different ways.

## Dynamic Lift

This lifting process results from the motion of the atmosphere at middle to upper levels in the troposphere. It is associated with the frontal lifting process in that the ascent results from the relative motion of air masses. The ascent may be more widespread than that associated with the immediate frontal surfaces. Generally, broad scale ascent is associated with the counter-clockwise motion around a low pressure system. Descent of air, or subsidence, is associated with clockwise flow around a high pressure system.

## Orographic Lift

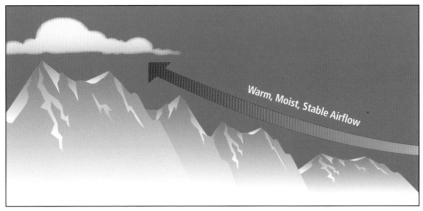

*Fig. 2-1* Moist air moving from east toward Rocky Mountains forms cloud

### Orographic Lift

Air can be forced to ascend as it moves along sloping terrain. This is known as orographic lift. Because of the presence of mountains and sloping plains, this process is very relevant in Alberta. It is predominant in the lower atmosphere, though if the mountains are high enough, as with the Rocky Mountains, the lifting process can extend well into the upper troposphere.

### Frontal Lift

A warmer air mass ascends over a colder, denser air mass. The interface between the two air masses is known as the frontal surface. One process that often governs ascent of air is associated with atmospheric frontal surfaces. This process is in many ways similar to orographic lift, except that the lifting agent is the frontal surface that surrounds the colder air rather than the earth's surface. Frontal surfaces can extend through the whole troposphere, so frontal lift can occur through a considerable depth.

## Frontal Lift

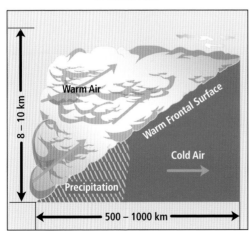

*Fig. 2-2* Warm air moving upward along a frontal surface

27

## Convection

One mechanism that produces ascent is convection. This process is common during daylight. As the sun heats the land surface, air adjacent to the land warms. The amount of warming depends on the type of land surface and its aspect ratio to the incident rays of the sun. Some air parcels are a little warmer than adjacent parcels and they rise. As they rise, they cool. If these parcels are warmer than the adjacent air at that level, they continue to rise. This process continues until the rising air parcels reach a level at which the surrounding air mass is warmer. At that level, the upward motion of the parcel ceases.

## Turbulent Lift

This mechanism is caused by the friction between a moving air mass and the underlying surface. Turbulent lift is somewhat similar to convection except that air parcels get their upward motion from undulations in the underlying surface. The height to which the turbulence extends depends on the speed of airflow (i.e., wind speed), the roughness of the underlying surface and the stability of the air mass itself. Turbulent lift generally does not extend far above the earth's surface, but it can be horizontally widespread.

### Convective Bubbles

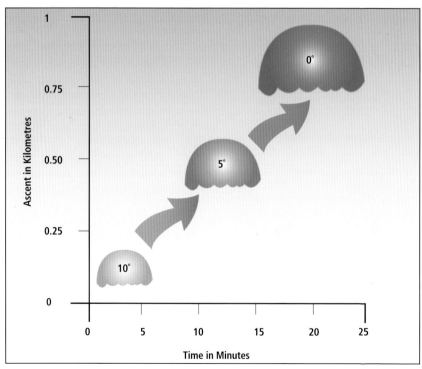

Fig. 2-3 Parcels of air rising as a result of convection

## Cloud Classification

An international classification system for clouds has been developed for use by ground observers. Clouds are generally classified into four families based on how high the cloud base is located above ground level. Low clouds have bases in the layer that reaches up to approximately 2 kilometres above ground level, though cloud based at ground level is generally called fog. Middle clouds are based between about 2 and 6 kilometres above ground level and are known as alto form (Latin for "middle"), whereas high clouds are based above 6 kilometres. Clouds that have a considerable vertical dimension and that are based in either the low or middle range are called cumulus (Latin for "heap").

Each cloud family is subdivided based on appearance. In general, the visual appearance of clouds is closely associated with the process that gives rise to the lifting process. Clouds that appear as a layer and are spread out horizontally are called stratus clouds (Latin for "layer"). Most high clouds are wispy in appearance, largely because they consist predominantly of ice crystals that are formed at the low temperatures associated with the upper troposphere. These clouds are known as cirrus (Latin for "curl of hair"). Latin terms are used to further define appearance (fractus, castellanus, pileus, lenticular, mammatus) or other cloud characteristics, such as the presence of rain. These terms are often used in combination. The nimbus subdivision is used for cloud types from which precipitation is falling.

What follows is a detailed description of the cloud types seen in Alberta.

## Stratus

As observed from the ground, these clouds appear as a uniform layer usually resulting in an overcast condition—that is, the sky is uniformly covered from horizon to horizon. These clouds are usually made up of liquid water droplets, but if the temperature at cloud level is cold enough, stratus clouds can contain a large number of ice crystals. Usually these clouds do not have any permanent markings on their lower surface, and individual cloud elements are indistinct. Precipitation from these clouds is rare and is at most a drizzle or light snow.

## Stratus Fractus

These clouds have a ragged appearance and are based at the same level in the atmosphere as a stratus deck. Stratus fractus clouds result when the conditions that formed a stratus cloud deck are no longer active and the stratus deck breaks up.

31

## Stratocumulus

Stratocumulus clouds consist of a series of patches of rounded clouds that have relatively little vertical development. These clouds are often associated with stratus clouds and are either a precursor to the formation of stratus or result from the transition of a stratus deck into cumulus cloud. The deck of stratocumulus clouds is not continuous; the edges are ill defined, and the cloud has a patchy or rolling appearance. Blue sky or higher cloud decks frequently appear through breaks in the layer of stratocumulus. Any precipitation associated with these clouds is at most a drizzle or light snow.

## Fog

Fog can be considered a cloud that is in contact with the ground. It can form when a moist air mass flows over a colder surface, or in calm wind conditions when moist air in contact with the ground cools overnight to the point that its relative humidity rises to 100 percent. When fog breaks up, usually in the presence of a slight breeze, a layer of stratus clouds may form above the ground. As ground heating increases in conjunction with an increase in surface wind, the base of the stratus deck may rise. Fog may also dissipate because of the action of sunlight, which heats the top of the fog layer. Some of the solar energy also penetrates the fog deck, warming the underlying ground surface and raising the air temperature beyond the point of saturation.

## Nimbostratus

As the nimbo preface implies, this cloud is often associated with continuous widespread rain or snow. This cloud is generally quite thick, extending from 1 to 2 kilometres above the ground at its base to tops often over 5 kilometres. The cloud base is usually dark grey to black, and cloud elements are indistinct.

## Helmholtz Waves

Helmholtz waves resemble breaking waves on water. These clouds are formed along a horizontal interface between warm air that is flowing over colder air. They are rarely seen because there is not usually enough moisture at that interface for clouds to form.

## Cloud Streets

Low clouds frequently align along the direction of low-level wind, resulting in what is known as a cloud street. These clouds are generally stratocumulus or cumulus. A steady horizontal wind initiates a rolling motion in the flow, with gentle ascent combined with gentle subsidence occurring along parallel lines. A line of cloud forms in the areas of ascent.

## Altostratus

Altostratus has a grey to steely blue colour (although not in this shot) and generally covers the whole sky. Thin layers of altostratus frequently resemble ground glass. This cloud has no characteristic thickness and can be up to 6 kilometres deep, which is thick enough to completely obscure the sun. The base of the cloud appears flat when observed from ground level, but when flying though an altostratus layer, the lower boundary is virtually indistinct. Precipitation rarely falls from these clouds.

## Altocumulus

This cloud type is similar to stratocumulus. The altocumulus cloud deck is not usually widespread. The cloud elements are arranged in groups, and they frequently form into lines, usually following the direction of the wind at their formation level.

### TYPE OF CONDITIONS CLOUDS USUALLY BRING:

**cirrus** — *usually means fine weather*

**cirrocumulus** — *the mackerel sky is sometimes an indication of unsettled weather*

**cirrostratus** — *approaching rain*

**altostratus** — *rain likely if cloud thickens*

**nimbostratus** — *rain or snow*

**stratus** — *maybe light rain or drizzle*

**cumulus** — *sunny days*

**cumulonimbus** — *thunderstorm clouds, showers, rain and sometimes hail*

## Altocumulus Castellanus
These clouds usually develop within a line of altocumulus. They are an indication that the middle atmosphere is unstable. When they form in the mornings during summer, they occasionally grow to be quite large vertically, and precipitation can fall from them. Occasionally they even develop into weak thunderstorms.

## Lenticular and Wave Clouds

These clouds are often observed in the vicinity of hills and mountain ranges, which makes Alberta a good place to see them. When air that is thermally stable vertically is forced to flow over obstructing terrain, it frequently forms a standing wave. Standing waves along the foothills of the Rocky Mountains are well known to glider pilots, who use them to provide the lift necessary to soar to heights above 10 kilometres. The moisture flowing through the wave pattern condenses in the updraft portion of the wave. Where the air subsides, cloud dissipates, giving the cloud the form of a curved lens. Because the waveform can often remain in position for a long time, the cloud is usually stationary in the sky for as long as the upper level winds and moisture conditions are favourable.

## Lee Wave Wind Pattern with Lenticular and Rotor Clouds

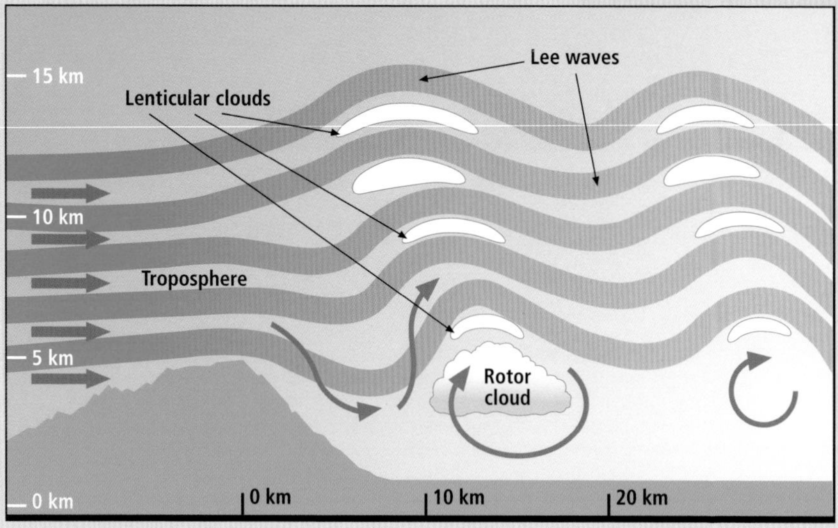

Fig. 2-4 A cross section of wind/flow associated with standing wave pattern

## Standing Waves and Chinook Arch

To the downwind (lee) side of the Rocky Mountains, standing waves can have a lengthy axis parallel to the mountain range. When looking toward the mountains from ground level, the clear sky framed by cloud to the east gives the curved appearance of an arch. This arch is often associated with surface winds flowing rapidly downslope from the west (a Chinook) and is dubbed the Chinook arch.

A cloud does not know why it moves in just such a direction and at such a speed...It feels an impulsion...this is the place to go now. But the sky knows the reasons and the patterns behind all clouds, and you will know, too, when you lift yourself high enough to see beyond horizons.

—Richard Bach

## Rotor Clouds

These clouds occasionally accompany wave clouds in the lee of ranges of hills. They are most frequently seen in the foothills of the Rockies during Chinook conditions. Precipitation does not usually accompany rotor clouds, and they are most notable for extreme air turbulence. They are therefore of great concern to low-level aircraft operation.

Standing lenticular cloud with rotor below

## THE COLOUR OF CLOUDS

*You can often tell what is going on inside a cloud by what colour it is. Storm clouds are dark because they have a high water droplet content and the droplets are tightly packed together, so light has trouble passing through them. Thunderstorm clouds take on a green tinge because light is scattered by the ice in the cloud. If you see a green-tinged cloud, you can be pretty sure that heavy rain and hail are on the way. A yellow cloud gets its colour from the presence of smoke, usually from forest fires. Yellow clouds are rare but can sometimes be seen during late spring or early fall, when forest fires are most common.*

## Cirrus

Cirrus clouds consist entirely of ice crystals. These clouds can take on a variety of forms but generally appear as fibrous wisps showing white against the blue sky. Isolated tufts occasionally have featherlike plumes that stretch out and turn upward. They are often called mares' tails.

## Cirrostratus

Also made up of ice crystals, cirrostratus clouds appear as a thin, whitish veil that can extend from horizon to horizon, giving the sky a milky appearance. Occasionally there is a fibrous appearance to the cirrostratus deck, which implies that there are streaks of thicker strands of cirrus cloud embedded within the deck.

Halos are sometimes seen around the sun or the moon as it shines through the cirrostratus deck. This phenomenon is often thought to be an omen of bad weather, which has some truth. A cirrostratus deck is formed from the overriding of moist air along an elevated frontal surface. This indicates that a different air mass is approaching, heralding a change in the weather.

45

**Cirrocumulus**
   These clouds have a patchy or, occasionally, wavelike appearance. They have a slight shadow, if any.

## Noctilucent Cloud

Noctilucent clouds occur in the mesopause at an altitude of about 80 kilometres. These clouds are made up of ice crystals, and the condensation nuclei are thought to be meteoric dust particles. The moisture content at this level is patchy, so the clouds are rarely observed. Because these ice crystal clouds are thin, they are only observed while the sun is still below the horizon. They have no effect on weather and are more of an observational curiosity. Alberta is one of the best places to observe these clouds.

## Contrails

Aircraft exhaust contains large amounts of water vapour and particulate matter. The exhaust gas cools quickly and, with the cold temperatures at flight level, usually results in the formation of ice crystal clouds that appear to stream out behind aircraft. The turbulent motion of the atmosphere causes these contrails to spread laterally, and the turbulence created by the aircraft means contrail dissipation usually occurs within a few minutes. However, if the atmosphere at the contrail level has a high relative humidity, the contrails may persist for a while longer. The horizontal motion of the atmosphere at aircraft flight level further complicates the phenomenon. While successive aircraft fly along the same path relative to the ground, the air at flight level is almost always moving. This gives the impression that successive contrails are side by side. The presence of successive and perhaps long-lived contrails may result in the formation of a thin cirrus shield. This shield will grow throughout the day, and in areas that are frequently crossed by over-flying aircraft, a fairly extensive deck of cirrostratus can result.

# Clouds with Vertical Development

The temperature of the ambient air that a parcel of rising air encounters governs how high a cloud can grow. If the temperature of the air within the parcel is warmer than the surrounding ambient air, the parcel continues to rise. When the air contained within a rising parcel becomes saturated, the latent heat energy is released. The cooling rate of the saturated air is reduced to about one-half of the cooling rate of the unsaturated parcel. Because the surrounding ambient atmosphere generally cools at a rate between the moist and dry rates, the rising saturated air parcel remains warmer than the surrounding ambient air. In this situation, the atmosphere is said to be convectively unstable, and clouds experience rapid vertical growth.

## Cumulus

These clouds generally have a flat base, and they are as high as they are wide. The edges of the clouds are usually quite distinct, giving the clouds the puffy, white appearance of popcorn. These fair weather cumulus clouds, often known as cumulus humilis, generally go through a standard life cycle. In late morning, the intensity of solar heating reaches a point where parcels of air near the surface rise, forming puffy cumulus clouds (we say the "cumulus starts popping"). As the day progresses, the appearance of the individual clouds slowly changes, and they may seem to be relatively motionless in the sky. As sun intensity decreases late in the day, the clouds generally dissipate. Precipitation rarely falls out of these cumulus clouds.

Towering cumulus clouds

## Towering Cumulus

The cumulus cloud base may be 1 to 2 kilometres above the ground, and the tops of towering cumulus may extend to heights of 4 to 5 kilometres. These clouds contain rather vigorous updrafts and often produce precipitation in the form of rain showers or snow flurries. Towering cumulus clouds may become more organized and can grow to a significant extent horizontally as well as vertically. These larger groups of towering cumulus are known as cumulus congestus.

Occasionally, the tops of towering cumulus will give an upward push to a layer of moist air aloft. This moist air condenses, forming a veil-like layer of ice crystal cloud (essentially cirrus). These veiled cumulus congestus clouds are known as cumulus pileus.

Towering cumulus clouds (above), cumulus congestus clouds (below)

## Cumulonimbus

The cumulonimbus cloud represents the ultimate growth phase of cumulus cloud into a vigorous, precipitating cloud mass. Cumulonimbus clouds can stretch horizontally for kilometres and extend from their base through the depth of the troposphere, to heights of 10 to 15 kilometres. Lower portions of these clouds contain liquid moisture (cloud and rain droplets), whereas upper portions consist of ice crystals. The upward growth of these clouds is constrained by the thermal stability of the tropopause. The tropopause acts as a lid, and the tops of cumulonimbus then spread laterally, forming a fibrous cirrus veil. Strong winds aloft generally move the veil downwind, resulting in the classic anvil-like appearance of the cumulonimbus cloud. The anvil top can extend many kilometres downwind of the main cloud base. This cloud almost always generates showery precipitation, and more vigorous storms produce hail within the cloud. This hail frequently reaches the ground.

## Fully Developed Cumulonimbus

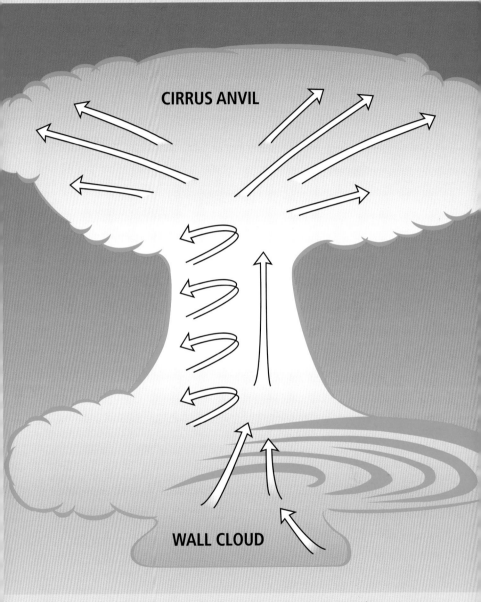

*Fig. 2-5* Air that spirals into the base of a cumulonimbus cloud can form a wall cloud. Air pushes out the top of the cloud to form a cirrus anvil.

# Chapter Three: Precipitation

Water vapour is present even in the driest of air masses, but it is the conversion of vapour into liquid or solid forms that creates much of the weather we experience. A simple analogy for how water vapour results in precipitation can be made with a sugar-saturated cup of hot coffee. When the solution cools, the liquid cannot continue to retain all of the sugar in its dissolved state, and some precipitates out as solid crystals.

Water vapour behaves like any other gas in the atmosphere, but atmospheric water has the ability to convert between solid, liquid and gaseous phases. The atmosphere cannot hold an unlimited amount of water vapour, and the amount of vapour that can be held is entirely

dependent on the temperature of the mixture. In the atmosphere, if the air at a given temperature is saturated with water vapour, cooling results in condensation into water droplets and sometimes sublimation directly into ice crystals.

Air does not contain a large amount of water vapour. One cubic metre of air at sea level, with a temperature of 35° C, weighs about 1 kilogram. If that cube is saturated with water vapour, the water has a weight of about 37 grams. We can measure atmospheric water vapour content in a number of different ways. Vapour pressure is the actual pressure exerted by water vapour, and it is always a small fraction of the air pressure. At a given temperature, there is a maximum vapour pressure that can be reached before a phase change occurs. This is known as the saturation vapour pressure.

Condensation in the atmosphere does not happen readily because the air molecules are always in motion. To stick together and form a liquid, the molecules require the presence of condensation nuclei. These nuclei are small, ranging from 0.1 to 10 microns (thousandths of a millimetre) in diameter. The lower atmosphere contains between ten thousand and one million condensation nuclei per cubic centimetre. In the upper atmosphere, there are hundreds per cubic centimetre. For ice crystals to

form, freezing nuclei are required. However, freezing nuclei are far less numerous than condensation nuclei. There may be only one or two per cubic centimetre.

Condensation forms cloud droplets that are typically 10 to 20 microns in diameter. When raindrop sizes are measured, they are typically found to be about 2000 microns in diameter. So, a typical raindrop is made up of more than one million cloud droplets. A major question for atmospheric scientists used to be what process occurs in the atmosphere to form raindrops from cloud droplets.

## Official Definitions of Precipitation

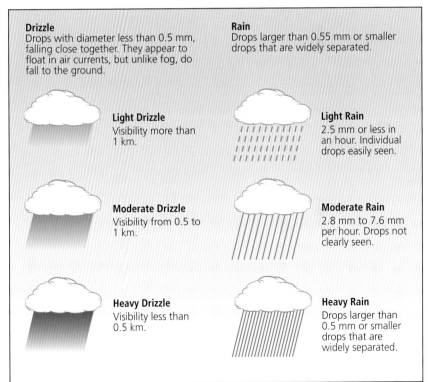

**Drizzle**
Drops with diameter less than 0.5 mm, falling close together. They appear to float in air currents, but unlike fog, do fall to the ground.

**Rain**
Drops larger than 0.55 mm or smaller drops that are widely separated.

**Light Drizzle**
Visibility more than 1 km.

**Light Rain**
2.5 mm or less in an hour. Individual drops easily seen.

**Moderate Drizzle**
Visibility from 0.5 to 1 km.

**Moderate Rain**
2.8 mm to 7.6 mm per hour. Drops not clearly seen.

**Heavy Drizzle**
Visibility less than 0.5 km.

**Heavy Rain**
Drops larger than 0.5 mm or smaller drops that are widely separated.

*Fig. 3-1*

# The Raindrop Formation Process

Precipitation forms in the presence of water vapour if cooling of air also happens. First, the cooling. For most cloud formation, this process usually occurs when a parcel of air is lifted. This lifting can happen when an air parcel is heated near ground level and becomes less dense than neighbouring air parcels. Another lifting process occurs when air moves over elevated terrain. Also, if an air mass moves horizontally against a cooler air mass, the air masses do not readily mix, and the warmer air rides over the colder air along the boundary between the air masses, which is called a front. This lifting process is known as frontal lift. The rate of cooling also depends on whether or not the air parcel is saturated with moisture. If the air parcel is not saturated, it cools at a rate of about 10° C per kilometre. A water-saturated parcel cools at a rate of 5° to 7° C per kilometre.

Two processes are thought to lead to the formation of raindrops. In the 1930s, scientists Alfred Wegener, Tor Bergeron and Walter Findeisen developed a theory about a process that is dependent on the presence of ice crystals. Ice crystals and water droplets form at temperatures below freezing, but when a mixture of water droplets and ice crystals is present, the ice grows more rapidly than the water droplets. In fact, water droplets do not spontaneously freeze until temperatures are below -30° C. So, at cold temperatures, ice crystals

spontaneously form and grow at the expense of water droplets. Gravity causes these ice crystals to fall. The cloud droplets also fall but at a much slower speed. The fall speed of droplets and ice crystals is dependent on their so-called terminal velocity. As they fall, the crystals continue to grow in the water-saturated air, and they collide with cloud droplets that have slower terminal velocity. These droplets immediately freeze when they strike the ice crystals, and splinters of ice may be ejected, which also grow into droplets. The ice crystals fall and, assuming that their fall speed is greater than the upward speed of the surrounding air, they move toward the ground. When they pass

## Growth of Ice Crystals

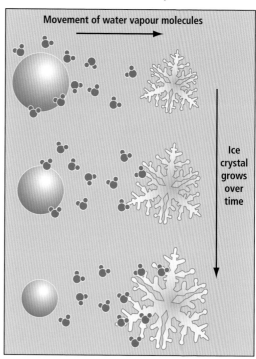

Fig. 3-2 Ice crystals grow more rapidly than water droplets when water vapour is abundant.

## Raindrop Growth by Accretion

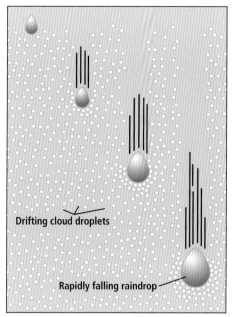

Drifting cloud droplets

Rapidly falling raindrop

*Fig. 3-3*

the rate of raindrop formation and the size of the resulting drops are directly proportional to the depth of the cloud. In Alberta, the precipitation formed in warm clouds usually consists of small drops that are known as drizzle.

Of course, the two precipitation-forming processes are usually underway at the same time in the atmosphere. In clouds that have significant depth, such as nimbostratus or cumulonimbus, both processes are happening within the cloud. If there are two distinct layers of cloud, snow and ice crystals that form in a layer with temperatures below freezing, such as cirrostratus, may fall into a lower layer of cloud that is, in part, above freezing. This snow can stimulate production of raindrops in the lower cloud, which normally may not be thick enough to form raindrops.

into air that is above freezing, they melt and form raindrops. This is the "ice crystal theory," also known as the "Bergeron process," which seems to explain the formation of snow and most rain.

The ice crystal theory, however, does not explain the formation of rain in clouds that do not contain below freezing temperatures. In the tropics, and even occasionally at mid-latitude locations such as Alberta, precipitation falls from warm clouds. A second theory known as "collision theory" explains this rain formation process. In this theory, the terminal velocity of different sized cloud droplets has to be considered. The larger cloud drops fall faster and overtake the smaller droplets, and the raindrops grow by accretion. With the warm cloud accretion process,

*A lot of rain starts off as snow, high in the atmosphere. It melts as it falls through warmer air near the ground, becoming rain.*

## Types of Hydrometeors

**Rain** is any liquid precipitation that reaches the ground. Raindrops can range in diameter from 0.5 to 7 millimetres. They are spherical in shape if they are less than about 2 millimetres in diameter, but larger raindrops falling through the atmosphere are flattened along the bottom because of the upward force of air. This force causes the breakup of drops larger than 7 millimetres across.

## Snow Shapes: Basic Structure

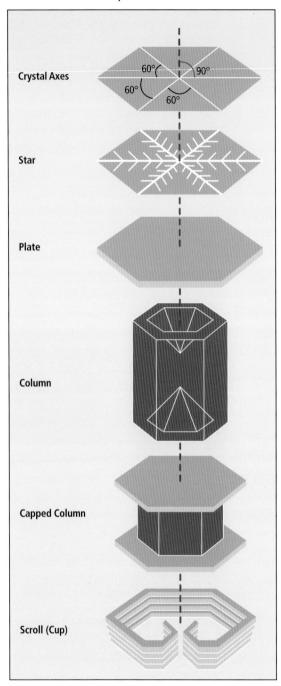

**Crystal Axes**

60° 90°
60° 60°

**Star**

**Plate**

**Column**

**Capped Column**

**Scroll (Cup)**

**Drizzle** consists of small drops that are less 0.5 millimetres in diameter. These drops fall out of relatively thin cloud decks and can be associated with fog. Drizzle drops have slow fall rates, and even the slightest breeze causes drizzle to appear to swirl. If temperatures are just below freezing at ground level, drizzle drops freeze on contact. The accumulation of ice from freezing drizzle can be significant and can result in treacherous travel conditions on Alberta highways. Freezing drizzle conditions are most frequent in Alberta during autumn.

**Snow** crystals come in a variety of shapes and sizes. Snow in the atmosphere forms shapes that are dependent on the nature of the water molecule. The water molecule, which consists of one oxygen and two hydrogen atoms, has a polarity based on the positive and negative charge distribution. When water molecules link together in an ice lattice, opposite electrical polarities of the atoms attract, forming a solid ice crystal that has a hexagonal symmetry. The hexagonal building block grows in one of many

*Fig. 3-4* The different shapes depend on the crystal structure of ice and the relative growth rates on each crystal axis.

characteristic fundamental shapes, which makes it possible to categorize snow crystals into several different classes. Figure 3-4 shows this crystal structure and the shapes of the various snow crystals and flakes than can result.

It is probably true that no two snowflakes are alike. As snow falls toward the ground, it can take on a variety of shapes. Perfectly symmetrical snow crystals are rare. Usually the air is at least somewhat turbulent, and the branches of stellar flakes can break off. Snow crystals may break as they strike each other, or they may stick together. Cloud droplets also strike the snowflakes and may freeze instantly, forming what is known as a rime coating. This causes the original crystal or flake to take on a lumpy, irregular shape.

As one would expect, average annual snowfall is highest over the high terrain to the west and lowest over the dry prairie lands in southeastern Alberta. Alberta takes the honour for the lowest annual provincial snowfall in Canada at 140.41 centimetres. Our province also has the

*Fig. 3-5* Temperature is the primary factor that controls shape, but available water vapour also has an influence.

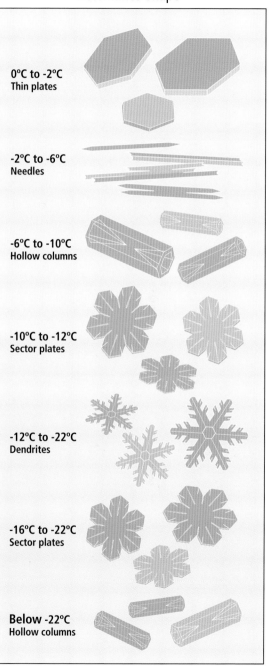

**Snow Shapes: Temperature Determines Shape**

**0°C to -2°C**
Thin plates

**-2°C to -6°C**
Needles

**-6°C to -10°C**
Hollow columns

**-10°C to -12°C**
Sector plates

**-12°C to -22°C**
Dendrites

**-16°C to -22°C**
Sector plates

**Below -22°C**
Hollow columns

## How Graupel and Hail Form

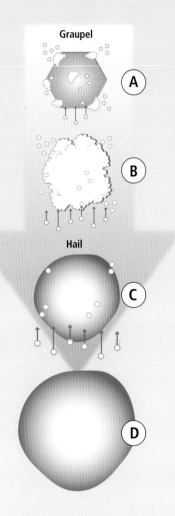

Fig. 3-6
A - supercooled cloud drops freeze to
   the crystal
B - cloud and raindrops freeze to form
   graupel
C - hail forms as more liquid raindrops
   freeze
D - hailstone falls out of the cloud when
   the updraft can no longer support its
   weight

most sunny days during the cold months at 114.61, giving some credence to the saying, "but it's a dry cold."

When ice crystals are subjected to a lengthy riming process as they fall through deep cloud, they form **graupel** or **snow pellets**. These heavily rimed ice particles are sometimes called soft hail, and they are usually no bigger than 5 millimetres. They may take on a conical shape. Snow pellets can be seen at ground level when temperatures are at or slightly below freezing. When they hit a hard surface, they may bounce but often shatter. Formation of graupel is important in the cloud electrification process. Graupel formation in deep convective clouds is the main mechanism for the formation of charged regions that lead to the generation of lightning.

### Graupel

Very large graupel samples compared to a dime, a nickel and a quarter.

## Formation of Ice Pellets and Freezing Rain

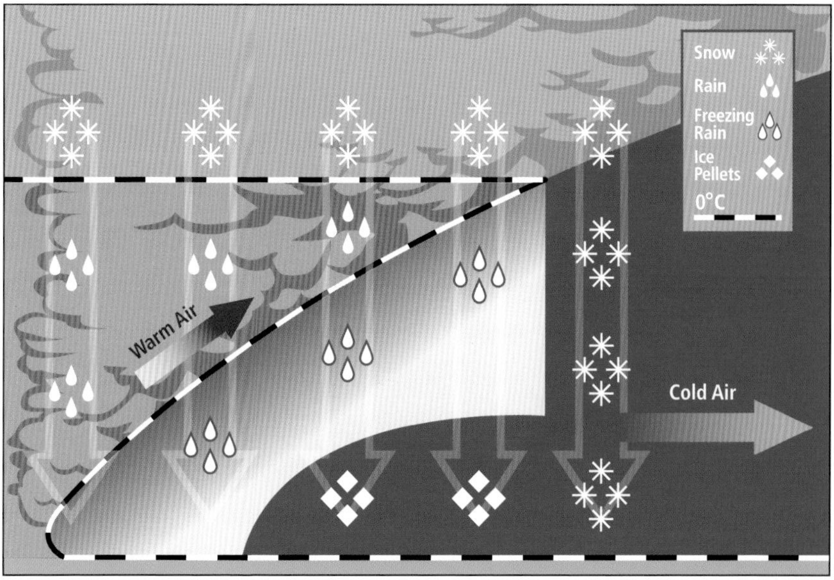

*Fig. 3-7* Freezing rain, ice pellets or snow depend on the thickness of the sub-zero air mass.

**Snow grains** are small, opaque particles of ice that consist of bundles of rime or ice crystals held together by frozen cloud droplets. They fall at a light, steady rate from layers of relatively thin cloud and are considered to be the solid equivalent of drizzle. They usually have diameters of less than 1 millimetre.

**Ice pellet** is the name commonly given to precipitation consisting of refrozen raindrops or refrozen, partially melted snowflakes. The particles of ice are usually transparent, or nearly so. Ice pellets as observed at ground level are formed when raindrops refreeze after falling through a layer of cold air near the ground surface. They are 1 to 5 millimetres in diameter.

When there is a significant layer of warm air aloft and a thin layer of cold,

sub-zero air at the ground surface, we get **freezing rain**. In this case, the surface-based cold air is not deep enough to freeze the falling raindrops, and they freeze when they strike objects at ground level that have a temperature below freezing. Such situations are relatively rare in Alberta, but when they happen, they can disrupt transportation. Airports, in particular, are keenly attuned to the possibility of freezing rain, because a coating on runways can shut down airports for lengthy periods if a prolonged cold spell follows a freezing rain event.

> Getting an inch of snow is like winning ten cents in the lottery.
>
> —Bill Watterson, *Calvin and Hobbes*

## Average Annual Number of Days with Freezing Precipitation

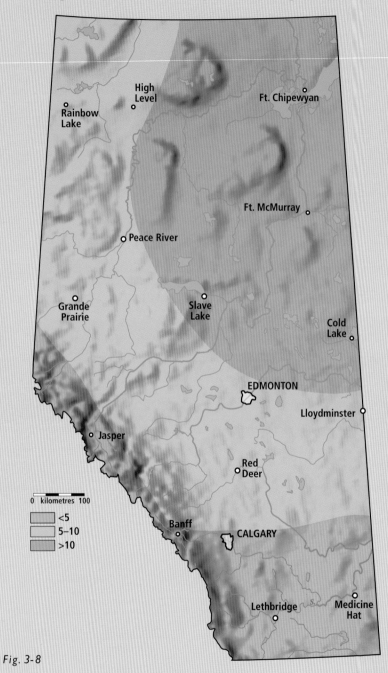

*Fig. 3-8*

Figure 3-8 shows the average annual number of days each year with freezing precipitation (rain or drizzle) in the province. The peak is in the northeast, with Fort McMurray averaging nine days per year and Cold Lake averaging 10 days per year.

**Hail** is the most destructive hydrometeor seen in Alberta. In their simplest configuration, hailstones are more or less spherical and are made up of concentric layers of clear and opaque ice. They form within actively growing convective shower clouds. Usually the electrical charge separation process is also underway in such clouds, so most hail occurs in association with lightning.

In active and developing convective clouds, strong upward and downward air movement can exist side by side. An individual hailstone begins its life as graupel that is supported by the strong updraft in the convective cloud. In strong updrafts, the graupel particle may rise a significant distance—perhaps several kilometres above its origin level. As it moves upward, it passes through regions that have high moisture content and a dense concentration of supercooled cloud droplets. With the high rate of collisions between the droplets and graupel, freezing is slowed for part of the accreted liquid, and the graupel becomes impregnated with water that freezes relatively slowly. The resulting

pellet is more or less transparent. The growth of a hailstone is illustrated in Figure 3-6.

The hailstone is buoyed by the strong updraft and can grow to a significant size as the water-saturated air flows past. In upper regions of the cloud, the updraft speed may reduce or the hailstone may move into an area with reduced uplift. When the stone becomes too heavy to be supported by the updraft, it begins a downward trajectory toward the earth's surface, all the while sweeping through supercooled droplets and continuing to grow. Often, the hailstone falls back into the stronger updraft and again moves upward. Figure 3-9 illustrates how hail growth occurs in a mature thunderstorm

## Structure of a Hailstorm

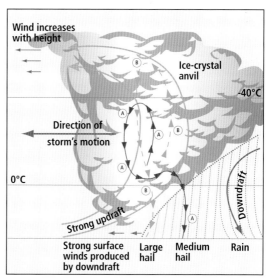

*Fig. 3-9*

A - hailstones grow in weaker updraft and fall out as medium-sized hail

B - large hail rises higher in cloud in strong updraft region

## Hailstone Cross Section

Different layers in a hailstone reveal different temperature and water vapour conditions at each growth stage.

cell. Updraft regions can be 1 kilometre or more across, and speeds in the most intense regions have been measured reaching 30 metres per second or more. The largest hailstones grow in these environments. Large hailstones often have a lumpy structure, possibly because the hailstones are spinning and tumbling in the updraft, and water streams off the stone, forming icicle-like projections. Many hailstones have an onion-like structure with many layers that formed as the stone moved between different updraft regions with different temperatures and liquid water content. This structure is revealed when cross-sections of hailstones are examined, as shown in the photo, above. Extremely large hailstones often consist of smaller individual stones that have collided and become frozen together within the storm cloud.

## Hail in Alberta

Hailstorms can occur anywhere across the province, but they have their greatest frequency in the region known as "hailstorm alley." This area located to the lee of the Rockies in central Alberta is subjected, on average, to more than six days with hail per summer. Other areas of relatively high hail frequency include the Peace Country of the northwest and the Cypress Hills in the southeastern corner of the province. Central Alberta experiences hail on 50 percent of days in the summer season, with larger hail (bigger than walnut size) occurring on 15 percent of days. An intensive study of storms in 1956 showed that hail occurred on 42 days during the summer within a 10,000 square kilometre area around Red Deer. Subsequent studies in the area showed that hail events are not sporadic events but are organized features. Whereas hail may last for only a few minutes at a specific location, the storms move along tracks that may be hundreds of kilometres long.

The 1987 Edmonton tornado also produced a lot of hail, with many reports of hailstones as large as tennis balls. The largest stone size reported was 17.8 centimetres along its maximum dimension. Damage from individual hailstorms can amount to hundreds of millions of dollars; for example, a hailstorm in Calgary in 1991 resulted in over $300 million dollars in insurance claims. Hail combined with heavy rain can cause flooding damage when storm sewers become

clogged with hailstones, preventing drainage (e.g., Calgary hailstorms of 1998, Edmonton hailstorm of 2003).

## Virga

This phenomenon deserves mention because it is often seen in dry environments such as that of Alberta. It is defined as streaks of rain or ice particles falling from the base of a cloud but evaporating before reaching the earth's surface. This phenomenon is frequently observed in Alberta because precipitating clouds are often based well above ground level, and the evaporation process is enhanced in the dry air at lower levels over the province. The presence of virga can be associated with dry downbursts, which themselves are a product of the cooling of air as it evaporates. In such cases, the presence of virga can indicate a hazard to aviation. Incidences of virga have also been misinterpreted as funnel clouds associated with thunderstorms.

Virga has the appearance of a funnel cloud.

# Chapter Four:
# Atmospheric Electricity

## Lightning

The atmosphere has both an electrical structure and a physical structure. In the lower portion of the atmosphere, lightning plays a key role as an electrostatic generator that helps recharge two concentric conductors—the surface of the earth and the electrosphere, which consist of highly conductive layers at altitudes above 50 to 60 kilometres. In the upper atmosphere, above 80 kilometres, the electrical phenomenon of most significance is the aurora borealis, which results when particles emitted by the sun interact with the atmospheric gases at those levels. There is not thought to be any interaction between the electrical processes in the lower and upper atmospheres.

The earth's surface has a negative electrical charge, whereas the lower atmosphere has a positive charge. This positive charge flows steadily to the earth's surface, in fair-weather conditions. In fact, if nothing else happened, the earth's charge would be neutralized in approximately 10 minutes, depending on the presence of conducting polluting gases. The thunderstorm acts as a generator that maintains the positive electrical charge in the atmosphere by means of the lightning discharge.

## Thunderstorm Generator Charges the Earth-Atmosphere Battery

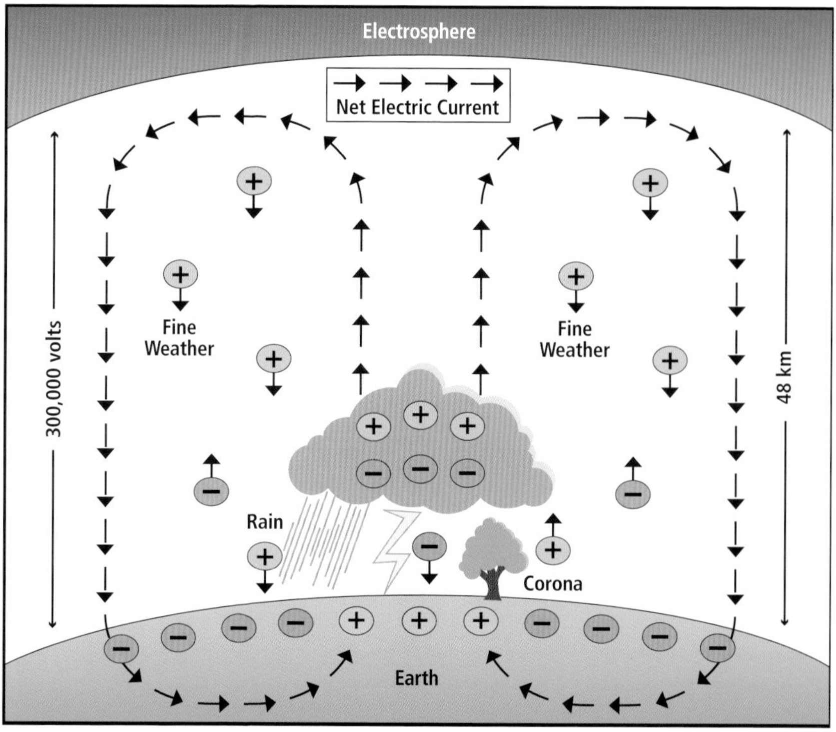

*Fig. 4-1* Thunderstorms act as generators to keep the earth charged negatively and the atmosphere charged positively.

Around the world there are approximately 2000 thunderstorms in progress at any given time, and with the rate of cloud-to-ground lightning flashes in an average thunderstorm, there may be 30 to 100 flashes to ground every second. In Alberta, the thunderstorm season is generally from April to October, though thunderstorms sometimes occur outside this period. During summer, there is usually at least one thunderstorm in progress every day.

*The Empire State Building in New York City has been hit by lightning up to 500 times in one year; it was once hit 12 times within 20 minutes. Toronto's CN Tower is hit 40 to 50 times per year. So much for "lightning never strikes the same place twice."*

## Convective Mechanism of Cloud Electrification

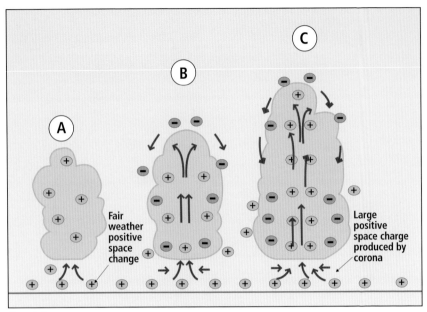

Fig. 4-2
A - positive charge swept into a developing cumulus
B - negative charge attracted toward edge of cloud
C - downdrafts on edge of cloud transfer negative charge to cloud base

# The Cloud Electrification Process

Two mechanisms are thought to be responsible for electrifying clouds, turning them into thunderstorms—the convection mechanism and the graupel-ice mechanism. Although the latter process is the most widely favoured, the two processes may work in tandem, and there may even be an electrification process that is still unrecognized.

## The Convective Process

Convective clouds go through a growth and intensification process that was described earlier. Convective clouds consist of updrafts of air parcels that are buoyant relative to the adjacent air mass.

These rising parcels contain moisture that condenses, forming the visible cumulus cloud. In the convection theory, shown in Figure 4-2, external sources provide the electrical charge.

*With lightning, the "leader" stroke is the one that reaches from the cloud to the ground, setting up a path for the "return" stroke, which travels from the ground up to the cloud. It is the return stroke that we actually see.*

Updrafts at the cloud base carry positive, fair-weather charges toward the top of a growing cumulus cloud, and the charge is more concentrated than that in the surrounding air. Negative charges produced by the constant bombardment of cosmic rays are attracted to the positive cloud and, in turn, attach to the cloud particles at the outer portions of the cloud. These negatively charged cloud parcels move downward in the convection circulation at the edge of and outside the cloud, compensating for the convective updrafts inside the cloud. This negative charge moves toward the cloud base where it can discharge to earth, producing a positive corona, which produces a positive charge under the cloud and a positive feedback process. This process does not explain all the complex charge distributions in fully developed thunderstorms.

## The Graupel-Ice Process

The graupel-ice process cannot function unless a precipitation process is present in a growing cloud. The precipitation elements are graupel particles. As gravity draws graupel particles toward the earth, they collide with cloud droplets and ice crystals that are at the same levels within the cloud. When collisions occur, the graupel particles, as well as nearby ice crystals, acquire either a positive or a

## The Graupel-Ice Charge Separation Process

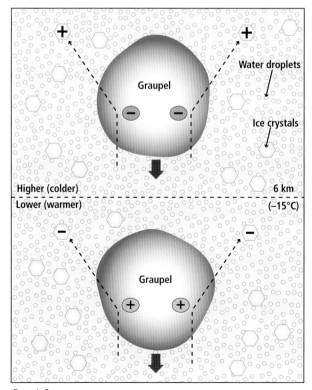

Fig. 4-3

## Charge Distribution in a Thunderstorm Cloud

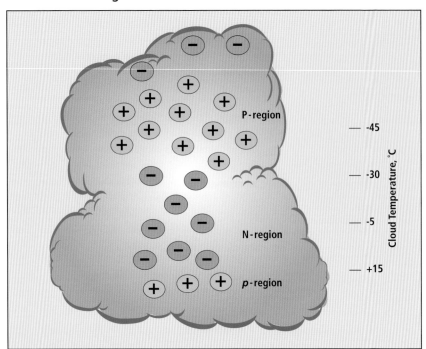

*Fig. 4-4* The likely charge distribution in a fully charged thunderstorm

negative charge. The sign of the charge depends on the temperature of the air, the water content of the cloud, ice crystal and water droplet sizes, the relative speed of particles in the collisions and the chemical pollutants in the water.

The most important factor is the temperature at which the collisions occur. In the upper part of the cloud where the temperatures are colder, the falling graupel acquires a negative charge and the crystals acquire a positive one. At warmer temperatures, the graupel is positively charged and the adjacent ice crystals are negatively charged. The critical temperature is between -10° C and -20° C.

This charge-separation process then results in what is known as the tripole distribution observed in thunderstorms (see Fig. 4-4). In this tripole, a positive charge is located at the top of the cloud (known as the P-region), a large region of negative charge is in the lower half of the cloud (known as the N-region) and a smaller region of positive charge (called the p-region) is at the cloud base.

*Lightning can heat the air it passes through up to 30,000° C. The surface of the sun has a temperature of about 6000° C.*

## Coronal Discharges and Lightning Forms

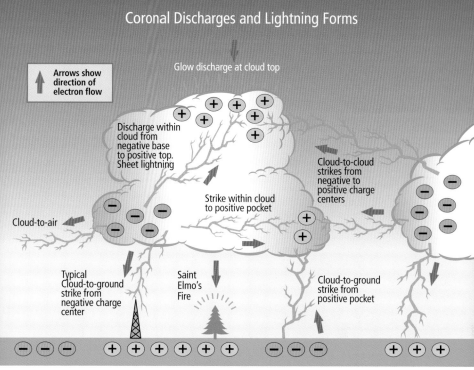

*Fig. 4-5* Saint Elmo's Fire is a coronal discharge that occurs when a thunderstorm is present. Many different types of lightning occur in thunderstorms.

# Generation of a Lightning Stroke

There are two primary types of lightning stroke: cloud-to-ground strokes (from the charged portion of the cloud to the earth) and intracloud strokes (between charged portions of a cloud). Occasionally, strokes pass between the charged portions of two clouds (called cloud-to-cloud lightning), and even more rare is a strike from cloud to air. During thunderstorms in North America, there are about five times as many intracloud discharges as cloud-to-ground discharges. Most of the pictures of lightning strikes are of the cloud-to-ground variety because the intracloud flashes are usually obscured by the cloud itself.

Although the charged cloud typically has a tripole configuration, the N-region is of greatest importance in cloud-to-ground discharges. As the cloud moves over the land surface, this negative charge region induces the buildup of a positive charge at the earth's surface. This positive charge area stays under the cloud as the cloud moves along.

The lightning stroke mechanism is complex, but most scientists believe the process starts when the negative and highly mobile electrons in the N-region discharge to the small p-region immediately below. The electrons overrun the p-region and continue their trip toward the ground in a rapid series of bursts along a channel known as the stepped

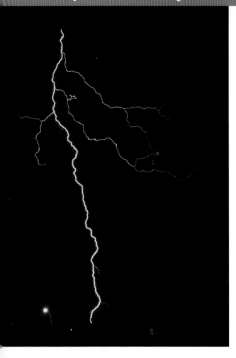

100,000 kilometres per second, or about one-third of the speed of light, and takes about 100 millionths of a second. This process is known as a single stroke discharge. The human eye is not capable of resolving the rapid speed of the luminous channel, and the flash appears to be one continuous stroke. Typically, however, several strokes occur within the same channel, progressively discharging higher and higher portions of the N-region. These consecutive strokes are typically separated by gaps of only 40 to 50 thousandths of a second. This rapid sequence of flashes gives lightning its flickering appearance.

The lightning discharge process is, of course, what is responsible for thunder. The rapid flow of electrons in the lightning channel causes a rapid heating of the channel to about 30,000° C. This heating causes an explosive expansion of the channel. This expansion compresses the surrounding air, producing a shockwave that propagates rapidly in all directions. After only a short distance, the shockwave converts to a sound wave that propagates at the speed of sound, which near sea level is about 330 metres per second.

leader. The lower end of the stepped leader moves toward the ground at about 120 kilometres per second. As the tip of the stepped leader approaches the ground, a strong positive charge is induced, especially on objects projecting above the ground surface. Upward-moving positive discharges, called streamers, are initiated from these points, and soon the downward stepped leader and the upward positive streamer connect. Electrons at the bottom of the stepped leader rapidly discharge to the ground and the lower portion of the channel becomes luminous. As electrons are pulled from higher and higher up the channel, the region of the strongest surge of downward rushing electrons also moves upward, and the associated luminous region moves upward.

The upward movement of the luminous region moves at a speed of about

## Anatomy of a Lightning Stroke

Fig. 4-6
A - electrons accumulate in the base of the cloud
B - electrons move downward along stepped leader
C - streamer moves upward from an object on the earth's surface
D - upward moving positive charge creates a luminous discharge channel

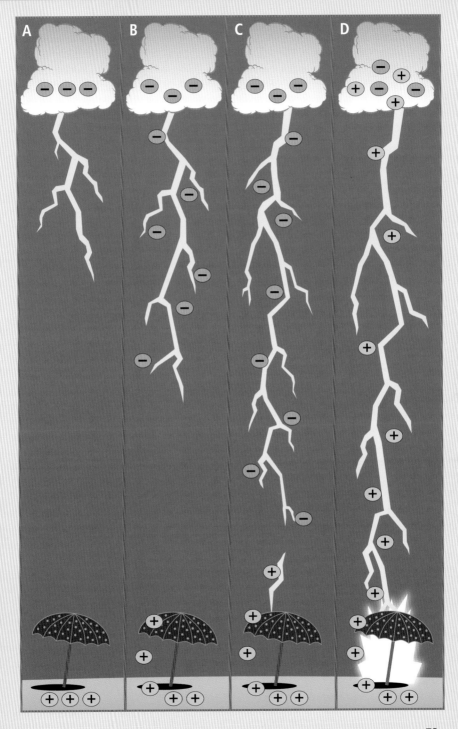

For an observer, the light from a flash appears instantaneous. However, the sound moves much slower. For a cloud-to-ground discharge, we hear the sound from the closest portion of the channel first and from the most distant portion last. Therefore, depending on the orientation of the lightning discharge channel, the thunder duration may be spread out over several seconds.

*The sound of thunder is a result of lightning rapidly heating the air it passes through, causing the air to expand explosively.*

The pitch of the thunder depends on the intensity of the explosive heating and the altitude at which the thunder shockwave is generated. To add further complexity, the atmosphere itself filters out higher pitched sound at a rate that is primarily dependent on the distance the sound has to travel through the air. Nearby thunder has a relatively high pitch, whereas more distant thunder has had a lot of the higher pitches suppressed and is a low rumble. In a nearby lightning discharge, we occasionally hear what sounds like hissing or cloth tearing. That sound is believed to be a combination of the sounds made by the downward moving stepped leader or the upward moving streamers.

The lightning process described has been concerned with discharges from the N-region in thunderstorms, but occasionally when the P-region is closest to ground level, it initiates a discharge. This can happen in a number of ways: if the cloud is strongly sheared such that the lower N-region is dispersed; if the cloud is relatively shallow and only develops a small N-region; or if the cloud is dissipating and the residual upper portion of the cloud remains intact. Thunderstorms forming in winter also have a higher frequency of positive lightning discharges because, in cold weather, the charge formation process favours establishment of a more extensive P-region. Lastly, a positive discharge can occur from a thunderstorm passing near a tall tower or elevated terrain, such as a nearby mountain peak.

Positive discharges typically consist of a single stroke, rather than the multiple strokes that are normally experienced. The positive strokes also carry a large amount of current to the ground. The magnitude of the total charge exchanges can be up to 10 times larger than those in a typical negative discharge. With these large current flows, positive strikes may also have a greater tendency to cause combustion and may be a primary cause of forest fires, especially because they often take place well away from a precipitating region of the storm cloud.

Although most lightning is of the cloud-to-ground or intracloud variety, upward discharge has also been observed from the top of the thunderstorm. There have been reports, mainly by high-flying U-2 aircraft, of upward propagating lightning strokes that have the same appearance as commonly observed lightning. These discharges appear to be a few kilometres in length.

Multiple strokes in a time-lapse photograph of a thunderstorm

> Thunder is good, thunder is
> impressive; but it is lightning
> that does the work.
>
> —Mark Twain

Other, more diffuse discharges have also been observed. Known as red sprites and blue jets, they extend 10 to 30 kilometres above large thunderstorms. They appear to occur in association with significant cloud-to-ground strikes. Although rare, their presence had been proposed as early as the 1920s. Sprites and jets are thought to form when a pulse of energy moves upward at the same time as a downward lightning discharge. This energy pulse, in turn, causes electrons to break free of gas molecules in the region above the thunderstorm. A cascade of energized electrons moves upward and causes the oxygen and nitrogen molecules to emit light. A ring of green light may also appear to flash at altitudes of 80 to 100 kilometres above ground level in association with the red sprite.

These phenomena are difficult to observe because clouds are usually blocking the view of a storm's top, and the actual discharge is faint. If they are seen at all in Alberta, they might be observed well past sundown with active thunderstorms that are 100 to 200 kilometres away. A possible scenario would be looking eastward in the late evening from an observation point in higher terrain (say the foothills) toward an active thunderstorm complex located near the Alberta-Saskatchewan border.

Ball lightning in a 19th century woodcut

## Ball Lightning

Small, luminous spheres sometimes observed during thunderstorms are called ball lightning. These balls are reported to be about the size of an orange or grapefruit, though some have been reported to be as large as 30 to 40 centimetres in diameter. The balls are usually located within a few metres of the ground. They move erratically, occasionally appearing to bounce along the surface. Lightning balls mostly appear yellow to red and seem to be rather innocuous. When they disappear after a short life span of seconds to one minute, they usually decay silently, though there have been reports of a small explosive sound. At present, there is no reliable theory for the formation and behaviour of ball lightning.

## Areas of Lightning in Alberta

Lightning detection networks, established to assist in forest fire detection and management programs, have made it possible to map the location and frequency of cloud-to-ground and intracloud lightning strikes. As shown in Figure 4-7, which plots the number of discharges per square kilometre per year, west-central Alberta is a lightning hot spot.

The number of strikes is, of course, closely related to the intensity of an individual thunderstorm. The more severe the storm, the more lightning discharges occur. One of the most intense storms recorded by the network was during the Edmonton tornado outbreak of July 31, 1987. On that day, it has been estimated that over 40,000 strikes were observed.

# Effects of Lightning Strikes

## Forest Fires

Given the global rate of cloud-to-ground strikes, forests worldwide might be expected to receive about 50,000 strikes per day. Not all strikes initiate forest fires, but the Alberta Department of Sustainable Resource Development estimates that lightning is responsible for starting half of the wild fires in Alberta. During the 10-year period ending in 2006, there was an average of 655 wild fires initiated by lightning. Studies of the characteristics of a typical cloud-to-ground stroke suggest that peak electrical currents are in the range of 10,000 to 20,000 amperes, with occasional peaks over 100,000 amperes. This current flows for a short period—usually less than one-thousandth of a second. Often there is a continuing current of 100 amperes or so that lasts for another one- or two-tenths of a second, and this type of stroke, with a short continuing current, is believed to be responsible for starting combustion. Lightning appears to follow a path through a thin layer of living cells between the inner bark and the wood layer. With the passage of the large currents through the tree, there is explosive heating—a strip of bark is generally blown off, and the tree itself may be split or splintered. Depending on the dryness of the tree bark and ground cover, a fire may be ignited. In many cases, the fire does not start immediately, and combustible material below ground level may smoulder for a few days before surface combustion begins.

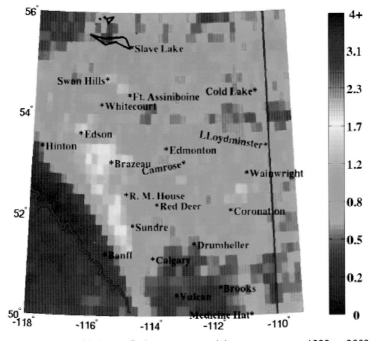

Fig. 4-7 The number of lightning flashes per square kilometre per year, 1998 to 2002

Fulgurites can form when lightning strikes sandy surfaces (below).

### Fulgurites

When lightning hits a land surface that is made up of sand and certain kinds of rock, the heat can be high enough to melt the material along the lightning channel. This material solidifies, creating what is known as a fulgurite. Fulgurites take the form of long, hollow tubes that range from 1 to 5 centimetres in diameter. They have been traced in surface sand layers to depths exceeding 15 metres. Ancient fulgurites have been uncovered that date back over 250 million years.

## Lightning Protection

Given the destructive nature of the cloud-to-ground lightning stroke, protecting property from lightning-initiated fire has a long history. Benjamin Franklin undertook experiments confirming that lightning had an electrical nature, and he was probably the first to suggest, in 1750, that lightning rods should be placed on structures to guide lightning to the ground through a conducting wire. Lightning codes have been established that outline the characteristics of an adequate protection system for residences. As well, high voltage electrical transmission lines have lightning protection systems built in. Combustible fluids must also be protected to ensure that lightning or any related electrical arcing does not contact any air-vapour mixtures.

*A lighting bolt can contain up to 100 million volts of electricity.*

# Passive Electrical Phenomena

There are a variety of other processes that occur in the atmosphere, the results of which can be observed in the sky in Alberta. These processes occur at various layers of the atmosphere and have a number of causes. They, for the most part, result in the production of a weak, light glow and can be classified as having an optio-electrical basis.

## Coronal Discharge

Coronal discharges occur any time there is a strong electrical field present in the atmosphere. Such strong electrical fields occur underneath active convective storms and cause ionization of the air molecules. When the gas is ionized, electrons in the molecule are elevated to a higher energy level. When the electrons return to their stable state, they emit light—a process known as fluorescence. This is similar to the mechanism that causes neon lights to glow. In the atmosphere, the coronal discharge appears as a luminous bright blue or violet glow that can be accompanied by a hissing or buzzing sound.

Electric fields are more concentrated in areas of high curvature, such as at the end of a lightning rod, ship mast, church spire, chimney or other pointed object. Aircraft flying through active electrical storms often develop coronal discharge streamers from antennas and propellers, and even from the entire fuselage and wing structure. The discharge can also appear on leaves and grass, from the tops of trees or thunderstorm clouds and even at the tips of cattle horns. Coronal discharge has also been associated with ships at sea during stormy weather. This is known as St. Elmo's fire.

## Lightning Safety Tips

**In the open**
- stay at least 30 metres away from metal fences
- remove shoes that have metal cleats
- do not use metal objects such as bicycles, golf clubs or fishing poles
- do not shelter under trees or canopies or in small sheds, picnic shelters and the like
- avoid open fields and high ground; if you are in an open field, crouch down and cover your ears
- seek shelter in low-lying areas, such as valleys or depressions, that are not prone to flash floods

**Indoors**
- keep windows and doors shut, and do not approach them
- do not have a bath or shower or use tap water—electricity can be carried through the pipes
- unplug all electrical appliances
- do not use a phone that is connected to a land line

**In a vehicle**
- you are safe in an all-metal vehicle as long as you do not touch anything on the interior that is metal
- do not park near trees or under power lines
- if a power line should fall on or near your vehicle, do not get out of the vehicle
- convertibles are NOT safe—it is a vehicle's outer body that makes it safe, not the rubber tires

79

## Aurora Borealis

All the electrical effects discussed to this point are associated with phenomena in the lower atmosphere. Albertans are also frequent witnesses to the aurora borealis, which are observed in the night sky of winter. The aurora borealis occur in the very high atmosphere, and though they have no apparent influence on the weather we experience at the earth's surface, they are sometimes put in a category known as space weather.

The earth has a strong magnetic field that is produced by the movement of the earth's molten core, and this magnetic field streams out into space. The sun is continually emitting a stream of positively and negatively charged particles known as plasma gas, which moves outward in what is often called a solar wind. Although the earth's magnetic field is greatly distorted by the pressure of the solar wind, the magnetic field serves the important function of keeping highly energized solar particles from reaching the earth's surface. The upper reaches of the earth's magnetic field also contain electrically charged plasma gas. The theory of the formation of the aurora borealis holds that the flow of the sun's plasma past the earth's plasma causes electrons to become highly energized. These electrons strike the rarified atmospheric gases more than 80 kilometres above the earth's surface. These collisions impart energy to, or excite, the gas molecules. When the excited gas molecules return to their stable state, they emit light energy, which is the source of the aurora borealis. The colours we see are characteristic of the types of gases that are present in the atmosphere and depend on the energy of the particles that are

stimulating the gas molecules. Atomic oxygen emits green and dark red light, whereas nitrogen emits blue and purple. The colours can be varied, and they shift depending on the motion of the magnetic field and the incident speed of the solar ions.

The shape of the auroral pattern is directly related to the shape of the magnetic field at that level. The aurora borealis generally have a fairly sharp lower cutoff at their base height of 80 to 100 kilometres. The top of the pattern is generally indistinct, but it can extend several hundreds of kilometres above the base.

The aurora borealis are quite spectacular and can often be seen in Alberta during winter. The frequency of aurora is greatest at about 60° N, so the frequency and intensity of auroral displays improve the farther north the observer is located in Alberta.

Another impressive lightning strike

It has been suggested that the aurora borealis emit sound; however, the region of excitation is located at least 80 kilometres above the surface, and the tone of the discharge has a fairly high frequency, based on what is observed with fluorescence discharge sounds in the lower atmosphere. Sound waves are preferentially absorbed by the atmosphere at higher frequencies, so it is doubtful that the sound would reach the earth's surface.

> **The weather is like the Government, always in the wrong.**
>
> —Jerome K. Jerome

## Air Glow

Air glow is a weak, blue-coloured emission by the atmosphere that forms as a result of several processes in the high atmosphere. During daylight hours, solar rays can cause molecules in the upper atmosphere to become ionized, and cosmic rays from all directions can excite atmospheric molecules. Chemical reactions between some gases present in the high atmosphere are also occurring. All of these processes result in the production of weak light energy. Although air glow is uniform across the sky, it is most readily observed toward the horizon because one is looking through a greater depth of the atmosphere, and therefore, the light is concentrated along those sight lines.

Photo courtesy of:TransAlta Energy Corporation

# Chapter Five: Winds

Pressure differences in the atmosphere set the air in motion, and temperature differences generally induce pressure differences in the horizontal. In the atmosphere, there are other forces that come into play on air parcels as well.

The rotation of the earth causes all parcels to be deflected. This is an apparent force, known as the Coriolis force. In the northern hemisphere, air parcels appear to be forced to the right, and south of the equator they are deflected to the left. This apparent force is strongest at the equator and decreases as you approach the north and south poles. In the absence of other forces, the Coriolis force causes the winds to blow along lines of constant pressure. In the northern hemisphere, this means that winds blow in a counter-clockwise direction around centres of low pressure and clockwise around centres of high pressure.

> **The pessimist complains about the wind, the optimist expects it to change; the realist adjusts the sails.**
>
> —William Arthur Ward

Any moving air parcel is also subjected to frictional forces, the most significant of which is the roughness of the underlying land surface. Frictional effects, which deflect winds by as much as 40 degrees toward lower pressure, are

83

Tree knock-down patterns in a wind storm

negligible above 400 metres over flat land or water surfaces and above 700 metres over rough terrain. Over water, the deflection is rarely more than 10 degrees. When looking at wind plots on maps of pressure at sea level, the drag of friction results in an apparent spiralling of winds toward the low pressure centres. The opposite is true for high pressure systems. Here friction appears to cause an outward spiralling of air away from the centre.

*Typhon, or Typhoeus, was the personification of intense windstorms, so a typhoon is a fierce storm.*

Curvature in the flow results in another force that affects the movement of air. This force opposes the motion of air along a curved path. It is generally a slight effect that results in the movement of air outward and away from the centre of low or high pressure.

The aforementioned forces modify large-scale circulation and wind patterns. Geographical features such as mountains, valleys and large lakes have a multitude of local effects on wind patterns. In addition, small-scale differences in heating are responsible for several wind patterns.

## Lake Breezes

Solar radiation is absorbed at different rates by land and water surfaces. Under sunny skies, air over land heats more quickly than air over an adjacent water surface. As daytime heating progresses, the difference between the air temperatures creates a corresponding pressure difference. The air over land has a lower pressure, and air over the water surface is pushed toward land by higher air pressure over the lake. The heated air rises, establishing a local circulation pattern, known as a lake breeze circulation, as shown in the diagram. The effect is most pronounced adjacent to large lakes, such as Lake Athabasca or Lesser Slave Lake, but it is present around all lakes. This circulation can also have a noticeable effect on the presence of clouds. Over land, the rising air parcels

## Lake Breeze Circulation

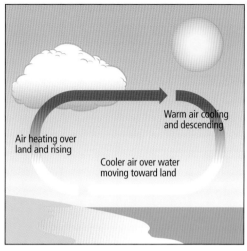

Air heating over land and rising

Warm air cooling and descending

Cooler air over water moving toward land

*Fig. 5-1*

may have enough moisture to condense and form small cumulus clouds, whereas subsiding air in the return flow over the lake is generally cloud free. And the circulation will progress during the day, strengthening and moving further inland. The flow pattern may also become visible if smoke is present and becomes entrained in the air circulation.

## Land Breeze Circulation

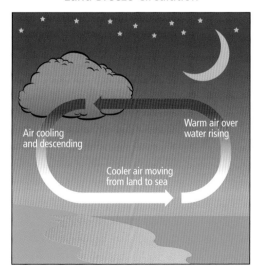

Air cooling and descending

Warm air over water rising

Cooler air moving from land to sea

*Fig. 5-2*

## Land Breezes

The opposite effect occurs at night when solar heating has ceased. The land surface cools more quickly than the adjacent lake surface, and a point may be reached where surface pressures are higher over land than over the lake. Air is pushed offshore, establishing what is known as a land breeze circulation. This circulation is usually not as pronounced as the lake breeze.

## Terrain Effect Winds

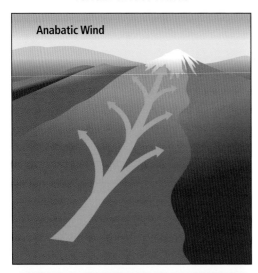

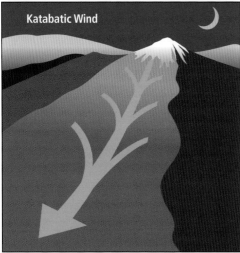

Fig. 5-3 Owing to heating and cooling, high terrain can substantially modify wind patterns.

# Terrain Effects

High terrain can substantially modify wind patterns, often distorting what may otherwise be a uniform pattern based on the larger scale pressure pattern. Thermal circulations that are established in valleys and near mountains are often observed in Alberta. During daylight hours, the sunlit side of a mountain may be strongly heated, and adjacent air will be significantly warmer than air further out from the mountain. The warmer air moves up the slope to higher elevations while the air away from the mountain subsides. This is called the anabatic wind pattern. At night, a reverse circulation is established. In this case, the mountain surface radiates heat and cools the adjacent air more quickly than the air farther out from the mountain slope. The cool air begins to sink and flows down the mountain surface, producing what is called a katabatic wind. This wind can be quite pronounced if the mountain surface is snow covered and cools adjacent air quickly. Katabatic winds can occur day or night in winter. They are usually much more pronounced than anabatic winds, and the mountain breezes that result can be quite cool.

*The trade winds get their name from the period in history when sailing ships were the main vehicles of trade. Captains of these ships factored in these winds when mapping out their journeys, sailing with the wind as much as possible.*

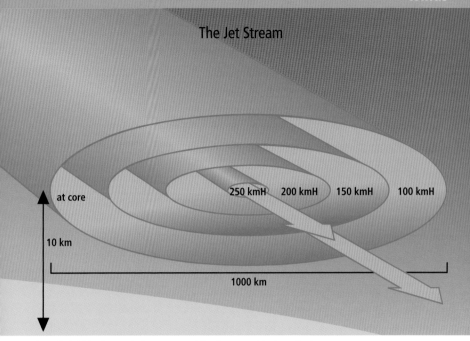

Fig. 5-4 Wind speed changes quickly above and below the jet stream core.

## Jets

Large temperature differences in the atmosphere give rise to strong winds called jets. There are two common types of jets—the jet streams in the high atmosphere, and the low level nocturnal jets that appear in the lower atmosphere.

The jet stream is located near the top of the troposphere, about 10 kilometres above the earth's surface. Jet streams at Alberta's latitude are usually associated with the frontal surface between air masses near the tropopause. At high altitudes, there may be several air masses, and therefore several jet streams, evident at any one time. Because the strength of the jet stream is a function of the temperature differences between air masses, the polar jet near the front that separates warm tropical air from colder, polar

maritime air is usually the strongest. And because colder air is poleward of warmer air, the wind within the jet stream usually blows from west to east.

Jet streams are truly weather phenomena with large linear dimension. When weather maps covering the whole hemisphere are examined, the jet streams may be seen to encircle the entire hemisphere. Where vigorous weather systems are present in the lower atmosphere, the jet stream appears to break into segments as short as 1000 kilometres, but often a jet stream extends to 10,000 kilometres. The width of a jet stream band can range from 400 to 1500 kilometres.

Jet stream wind speeds are usually stronger in winter than in summer because of the greater contrast between warm and cold air masses. The jet stream

## Jet Stream Positions

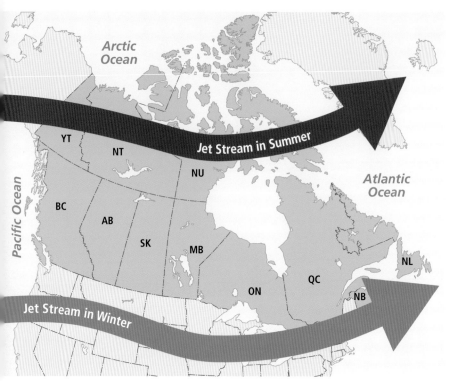

*Fig. 5-5* The jet stream's average position moves north in summer.

band will vary greatly in width. The wind speed peaks at the core of the jet and can reach speeds of over 400 kilometres per hour, but the maximums are generally in the 150 to 200 kilometre per hour range. The mean position of the jet stream also moves with the seasons. In winter, the strongest jet stream is usually positioned south of Alberta. In summer, the strongest jet stream cores may be located through or even north of Alberta for short periods.

For most of us, the importance of the jet stream is how it relates to air travel. Aircraft flying in the same direction as the jet stream have their speed enhanced by the same amount as the wind at flight level. So an aircraft flying across the continent from west to east may have a flight time that is 1½ hours shorter than an equivalent flight from east to west. Usually then, airlines prefer to fly on routes that take the jet stream into account.

The downside of flying the jet stream is the turbulence that is often associated. Turbulent flow in the atmosphere occurs when winds change speed rapidly. Jet stream flow is not horizontally uniform, and the speed drops more quickly poleward of the jet maximum than equatorward. As well, the wind drops quite quickly into the stratosphere above the jet core. So flying the jet stream is a balancing act between travel time and turbulence, and commercial airline pilots are constantly monitoring the locations of jet streams and associated areas of clear air turbulence.

> **Remember, the storm is a good opportunity for the pine and the cypress to show their strength and their flexibility.**
>
> —Ho Chi Minh

## Nocturnal Jets

At night as the ground surface cools, so does the adjacent air. The air above the surface layer does not cool as quickly, so a warmer layer of air may be above the colder surface air. This situation is reversed from the normal change of temperature with height, in which air cools with elevation. The reversed lapse rate condition is known as an inversion. Nocturnal inversion conditions occur most nights. This thermal structure in the lower atmosphere often induces weakening of the wind adjacent to the ground surface, so the winds we experience at ground level are light over night. At the top of the nocturnal inversion, the winds reach a maximum, which is known as a nocturnal jet. Maximum wind speeds can reach over 60 kilometres per hour. The depth of this nocturnal jet is only 100 metres or so, but because the nocturnal inversion is widespread and quite uniform over the prairies, the nocturnal jet may exist as a "sheet" of strong winds

## Winds from Around the World

**Brickfielder:** a strong, dry, dust-laden summer wind from the desert in southern Australia.

**Haboob:** a strong wind that occurs in the Middle East and along the southern edges of the Sahara in the Sudan; is associated with extreme sandstorms and, occasionally, thunderstorms, and is most common in summer.

**Levanter:** a strong easterly wind in the Mediterranean, especially the Strait of Gibraltar; is associated with foggy, overcast or rainy weather, especially in winter.

**Nor'easter:** a strong northeast wind that blows across the Atlantic provinces and the coast of New England.

**Pampero:** a strong west or southwest wind in southern Argentina that brings with it cold polar air and is often associated with severe thunderstorms.

**Santa Ana:** a strong, hot, dry wind that blows from the southern California desert through the Santa Ana Pass; most common in spring and autumn, but can form at any time.

**Simoom:** means "the poisoner"; a strong, dust-laden, cyclonic wind that blows across the Sahara Desert and areas of the Arabian peninsula, including Jordan and Syria, bringing with it temperatures in excess of 54° C.

**Sirocco:** a hot wind in southern Spain that originates in the Sahara Desert.

**Taku:** a powerful northeasterly wind that occurs occasionally in Alaska between October and April; winds speeds can reach hurricane levels.

several hundreds of kilometres wide and up to 1000 kilometres long. This jet breaks up in the morning as the nocturnal inversion dissipates with heating. On occasion, the jet becomes so intense overnight that the stability of the atmosphere breaks down. In this case, surface winds may become gusty under clear sky conditions. This is usually accompanied by a temperature increase as the warm air aloft mixes to the surface. Usually the gusty conditions are not prolonged and the lower atmosphere repeats the inversion formation process. I have observed this phenomenon while working overnight as a shift forecaster—winds suddenly increased at a few locations for no "apparent" reason (i.e., no fronts or thunderstorms in the vicinity). It can be quite unsettling when the tranquillity of a clear, calm night is shattered.

### The Tower of Wind in Athens

*Built in 100 BC, the tower is an eight-sided structure, with each side depicting one of the main winds recognized by the ancient Greeks. The Greeks personified the winds, giving them personalities and attire that represented the type of weather each one brought. The eight winds recognized by the Greeks were Boreus (north), Notos (south), Zephyros (west), Apeliotes (east), Kaikas (northeast), Euros (southeast), Lips (southwest) and Skiros (northwest).*

## Convection-Induced Winds

Cumulus clouds are formed when air near the ground is heated and begins to rise in features known as thermals. As the parcels rise, other air parcels move downward to replace the rising air. These parcels mix with the air at lower levels, resulting in a turbulent layer of air below the cloud base. An observer on the ground would feel the movement of this turbulent air as a gusty wind.

As cumulus clouds develop vertically, precipitation may start to form within the cloud. The raindrops start to fall relative to the uplifting air, and as the drops grow, their downward velocity increases until it is faster than the upward moving air. This rain falls through the cloud, and when it exits the saturated base, evaporation starts. The evaporation process extracts heat from the surrounding air and may cause cooling. The air within the rain shaft may, in fact, become colder than surrounding air, causing an accelerated downward push. The downward rushing rain mass also drags the air immediately adjacent to the droplets so that an extensive slug of cold, moist air accelerates toward the ground. Upon reaching the underlying land surface, this downdraft spreads out laterally, pushing dryer air ahead. At the ground surface, we observe this as a gust front. The gust front may be located well downwind of the precipitating shower cell, and it can appear in any direction from the cloud. When a gust front associated with a precipitating rain cloud passes, there is sometimes a sudden wind shift with no precipitation. Most often,

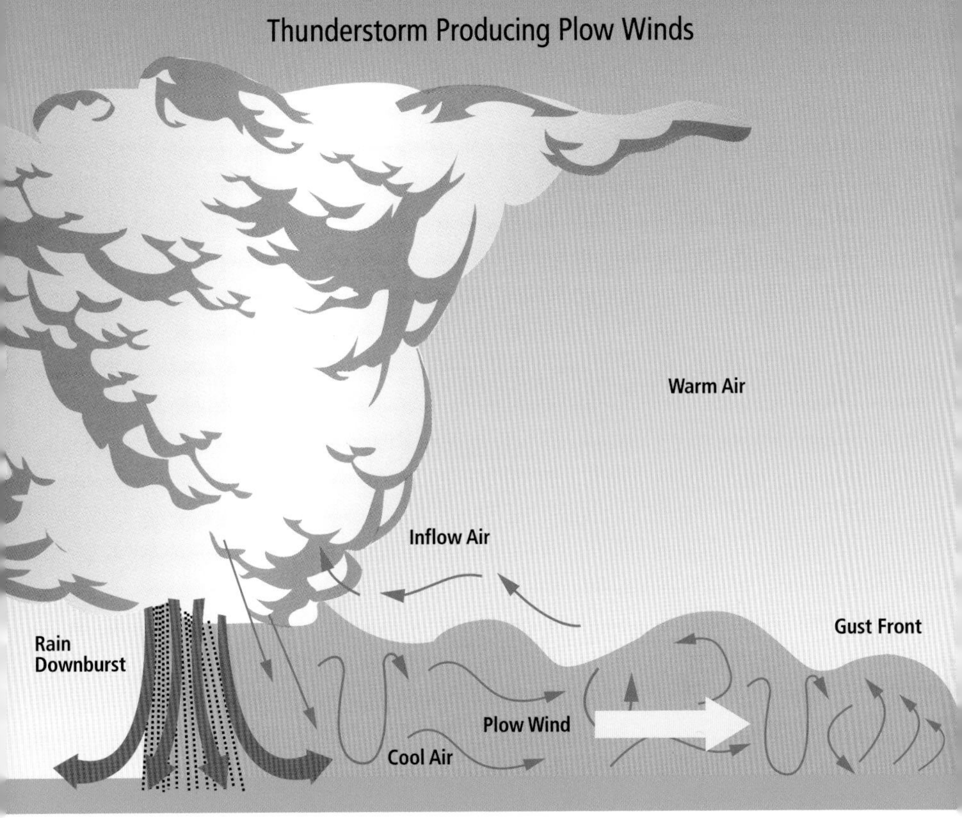

## Thunderstorm Producing Plow Winds

Warm Air

Inflow Air

Gust Front

Rain
Downburst

Plow Wind

Cool Air

*Fig. 5-6* Downward-rushing cold air moves rapidly along the ground, producing a plow wind close to the base of the storm and a gust front.

however, the wind shift is soon followed by rain. This gust front has a number of impacts. The cold air can inhibit further convection near the raining parent cloud by choking off updrafts. The gust front can also undercut warm, moist air that is moving toward the parent cloud. Occasionally, the uplifted, moist air at the gust front starts a new line of convective showers. Intersecting gust fronts from thunderstorms with vigorous rain can result in the formation of a new rain cell, which may in turn grow into a mature thunderstorm.

The downburst from an actively precipitating thunderstorm occasionally results in strong winds that may damage trees or structures in their path. Wind speeds in severe downbursts can exceed speeds of 250 kilometres per hour, the same speeds that are observed in tornadoes. In the Canadian prairie provinces, these destructive downbursts are often called plow winds because they cause damage that is generally in a straight line, generating a destruction path that is akin to the passage of a plow in soil.

## Tornadoes

The winds associated with tornadoes are the strongest in the atmosphere. Conventional wind measurement instrumentation cannot withstand the force of these winds, so the strength is usually an estimate based on the type of damage that occurs. The speed scale is named after Theodore Fujita, the research scientist who developed it in the 1960s. The so-called F scale describes wind speeds in six ranges, from F0, with wind speeds up to 115 kilometres per hour, to F5 tornadoes, with wind speeds estimated over 415 kilometres per hour. In Alberta, few tornadoes have speeds above the F3 level.

The large tornadoes associated with the Edmonton storm on July 31, 1987, apparently reached the F4 level with wind speeds of 330 to 400 kilometres per hour, whereas the Pine Lake storm of July 14, 2000, was estimated at F3.

The tornado formation process is still not fully understood, but observations of the tornado storm environment together with the understanding of thunderstorm dynamics are providing a credible theory. The majority of thunderstorms go through a normal lifecycle, from formation to decline, within about one hour. A few cells, known as supercells, last for several hours and can track several hundreds of kilometres across the prairies. Some supercells develop into what are called mesocyclones, with which the stronger tornadoes are associated.

Updrafts and downdrafts are associated with thunderstorms. As the thunderstorm cell moves forward, low level winds feed moisture into the cell below its base. This moist air condenses, releases latent heat and creates air parcels that can accelerate upward in areas known as updrafts. The speed in the updrafts can be more than 30 metres per second. Downdrafts composed of down-rushing, cold, precipitation-laden air can have speeds in excess of 10 metres per second. These updrafts and downdrafts are often adjacent to each other, and the zone between them is subjected to a high wind shear. This wind shear can impart a horizontal rolling action to the atmosphere just above the ground, near the cloud base. This rolling wind tube sometimes becomes vertically tilted, and if the rotation is cyclonic (counter-clockwise) it may then impart a cyclonic circulation to the thunderstorm, creating a mesocyclone. As the mesocyclone continues its movement, ingesting warm, moist air and generating downdrafts, roll clouds continue to form and may be tilted onto a vertical axis. These rotating tubes stretch upward with the updraft, and if they have a counter-clockwise rotation, they may link with the larger

## Fujita Scale Rating the Severity of Tornadoes

**The Fujita scale is used to rate the severity of tornadoes as a measure of the damage they cause.**

| INTENSITY | ESTIMATED WIND SPEED | DAMAGE |
|---|---|---|
| F0 | light winds of 64–116 km/h | some damage to chimneys, TV antennas, roof shingles, trees, signs and windows |
| F1 | moderate winds of 117–180 km/h | cars overturned, carports destroyed and trees uprooted |
| F2 | considerable winds of 181–252 km/h | sheds and outbuildings demolished, roofs blown off homes, and mobile homes overturned |
| F3 | severe winds of 253–330 km/h | exterior walls and roofs blown off homes, metal buildings collapsed or severely damaged, and forests and farmland flattened |
| F4 | devastating winds of 331–417 km/h | few walls, if any, left standing in well-built homes; large steel and concrete missiles thrown great distances |
| F5 | incredible winds of 418–509 km/h | homes levelled or carried great distances, tremendous damage to large structures such as schools and motels, and exterior walls and roofs can be torn off |

mesocyclone. The stretching process further enhances the rotation speed of the tubes, much like how figure skaters increase their rotation speed by pulling their arms closer to their body. At this point, the spinning column becomes a tornado. The most severe tornadoes often have smaller spinning vortices that move around the main rotating funnel. These are known as suction vortices and frequently contain the strongest winds. Suction vortices can be observed in video footage of the Edmonton tornado.

*If you are outside when a tornado touches down, the safest place to be is in a ditch.*

The length of time a tornado is on the ground is highly variable. The parent storm complex often generates a series of tornadoes. The 1987 Edmonton storm appears to have consisted of two tornadoes, the first of which touched down near Leduc and moved along the ground for a few kilometres. That tornado retracted into the parent cloud, and the second, most destructive funnel formed somewhat to the east of the first funnel. The total path length along the ground of the tornado funnels was 41 kilometres over a period of about 60 minutes. The Pine Lake tornado was apparently a single funnel that had a path length of 20 kilometres and was on the ground for about 35 minutes. A long-lived storm complex was associated with a tornado that touched down near Athabasca in 1984. The associated thunderstorms formed near Drayton Valley in mid-afternoon and tracked toward Stony Plain, where a funnel cloud was observed around 4 PM. The storm continued to track to the northeast, and another funnel developed in mid-evening, causing damage and one death over an intermittent path of about 40 kilometres. The storm continued tracking northeast over a largely unpopulated area toward Fort McMurray. Near midnight, a lightning strike from the still-vigorous thunderstorm caused the death of a Fort McMurray resident. The total track length of this storm complex was more than 400 kilometres.

## Formation of a Tornado

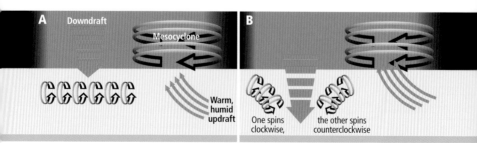

*Fig. 5-7*
A - low-level wind shear forms a spinning tube of air
B - the storm's downdraft pushes down on the spinning tube, tilting it into two columns

Another example of a tornado (above).

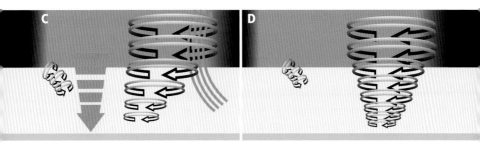

C - the counter-clockwise column is stretched into a tornado by the updraft

D - the column spinning clockwise, unstretched, spins slower. Sometimes it circles the tornado as a small funnel cloud.

## Top Alberta Tornadoes

| Number | Location | Date/time | Severity (Fujita Scale) | Anecdotal |
|--------|----------|-----------|-------------------------|-----------|
| 1 | Edmonton | July 31, 1987 3:30 to 4:30 PM | F3–F4 | 27 deaths; 200+ injuries; $250M in damage claims |
| 2 | Pine Lake | July 15, 2000 6:00 to 7:00 PM | F2–F3 | 15 deaths; 130+ injuries; $12M in damage claims |
| 3 | Grassy Lake | June 26, 1915 7:00 PM | F4 | 5 deaths; 9 injuries, train derailment; possibly 4 tornadoes along a 40 km path |
| 4 | Stoney Indian Reserve / Calgary | August 12, 1950 6:00 PM | F2 | 4 deaths, all from one demolished log house; 6 injuries; 15-metre trees uprooted |
| 5 | Wainwright | July 31, 1918 8:00 PM | F3 | 4 deaths; 3 injuries |
| 6 | Big Coulee / Athabasca | June 29, 1984 4:30 to 6:00 PM | F3 | 1 death; 1 injury; 35 farms struck in series of tornadoes |
| 7 | Okotoks | May 23, 1925 | F1 | 4 injuries |
| 8 | Wetaskiwin / Brightview | July 9, 1927 12:30 AM | F1 | 3 people killed while sleeping in granary |
| 9 | Redcliff | June 25, 1915 | F3 | 5 injuries; 9 businesses destroyed |
| 10 | Bawlf | July 28, 1972 | F3 | 1 death; 2 injuries; farmhouse destroyed |
| 11 | Esther | June 29, 1912 | F0 | 1 death; same day as the Regina Tornado |
| 12 | Travers Dam Trailer Park in Lethbridge | August 4, 1960 6:15 PM | F2 | 1 killed; 3 injured in trailer park |
| 13 | Lloydminster | July 9, 1983 | F3 | Homes and farm buildings demolished, financial aid given because of livestock losses |

*Fig. 5-8*

## Dust Devils

Dust devils are basically weak cousins of tornadoes. They are not associated with any cloud and are formed during late spring and summer on nearly calm, sunny days by the heating caused by intense isolation on a dry or sandy terrain. Air in contact with such surfaces can become super heated. Hot air rises rapidly in a column above the hottest area of the terrain, and hot air swirls in at the base of the column, so that the whole column takes on either a clockwise or counter-clockwise rotation. Dust devils are made visible by the dust, sand or debris that they pick up from the ground. The most vigorous dust devils can extend to heights approaching 1 kilometre, though most are only a few metres in depth. Surface wind strength can be up to 70 kilometres per hour. Dust devils are usually short-lived and move slowly in the direction of the mean surface wind.

A dust devil forms on a hot, sunny day with no associated cloud (opposite page).

Cold air funnels can hang below intense convective rain clouds.

Fanatics in power and the funnel of a tornado have this in common—the narrow path in which they move is marked by violence and destruction.

—Oscar Ostlund

Many funnels are long and slender (opposite page).

## Cold Air Funnels

These weather phenomena are small funnel clouds that rarely reach the earth's surface. When they do touch ground, and by definition become tornadoes, they are short lived. Their strength is usually in the F0 category and they cause little damage, if any. In Alberta, vigorous developing convective rain cells in spring and early summer produce these cold air funnels, and there are many reports of cold air funnel sightings every spring.

# Windstorms and Tornadoes

## Tornado Watches

- Issued where there is significant potential for cold core funnels, waterspouts or rotating thunderstorms that may spawn tornados
- Can be issued for an individual region or a group of regions
- Lead time up to 2 hours, but usually much shorter
- There is time for preparation (secure loose items outdoors, notify others)
- Inform public to watch the sky for developments

## Tornado Warnings

- Issued when a tornado is expected to develop soon, a tornado is nearby and will move into the area or a tornado is in the area
- Specific to a region or an area within a region
- Lead time up to 10 minutes but can be shorter
- Take cover

## Windstorm and Tornado Safety

### Outdoors

- Do not chase storms or tornadoes—they can change direction unexpectedly
- Get out of the storm path by moving at right angles to the direction of the storm's motion
- Get out of mobile homes or campers and go to a sturdy, permanent building
- If driving, special care and attention is required
  - obey traffic laws
  - be wary of other drivers who may panic
  - do not seek shelter under a road overpass

### Indoors

- Go to the lower floors, such as a basement under stairs
- Put as many walls as possible between you and outdoors
- Avoid windowed walls
- Washrooms, hallways, stairwells and reinforced walls are best
- Avoid large open-span roofed areas such as gyms
- Hide under heavy desks, tables or workbenches

*Fig. 5-9*

Chinook arch as viewed in southern Alberta

## Chinooks

These winds are formed along the lee slopes of the Rocky Mountains and are the best known of the mountain-induced circulations. Chinooks occur when Pacific air warms as it descends from the Continental Divide, scouring out the colder arctic air that lies in the lee of the Rocky Mountains. The Chinook (from the Salish Indian word *Tsinuk*, for "snow eater") is accompanied by the sudden onset of strong winds and warm temperatures. Because the air mass in the down-rushing air has lost most of its moisture during the climb from near sea level on the west coast, Chinook winds are very dry and have a desiccating effect along the lee of the Rockies. Chinook winds can be quite strong and gusty, with speeds often in excess of 100 kilometres per hour. When Chinook winds are funnelled through narrow valleys aligned from east to west, even stronger winds can occur, occasionally upward of 160 kilometres per hour. These winds are strong enough to force large vehicles, such as transport trucks and motorhomes, off the road.

The high frequency of Chinooks and strong westerly winds in southern Alberta makes the area around Pincher Creek an excellent location for wind farms. Windmills have been located along ridges in the Crowsnest valley to take advantage of the steady westerly flow. The most recently opened wind farm near Taber has 37 wind turbines, which generate 80 megawats of electrical energy that is fed into the Alberta electric power grid. Other locations across Alberta where Chinook winds predominate are being considered for wind farming as well.

*In Greek mythology, Zephrus was the personification of the west wind, so a zephyr is a gentle breeze.*

## The Formation of Chinook Winds

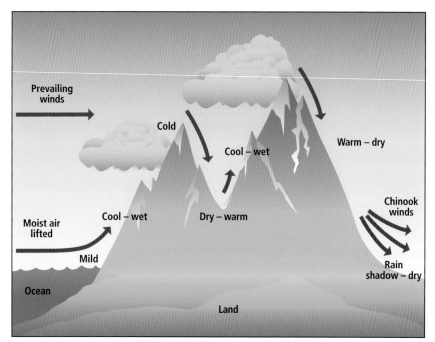

*Fig. 5-10* Moist air is forced to rise over a succession of mountain ranges. The Chinook winds sweep down the lee of the Rockies into Alberta.

Satellite photograph showing Chinook (white clouds) forming east of the Rockies.

# Average Annual Number of Days with Chinooks

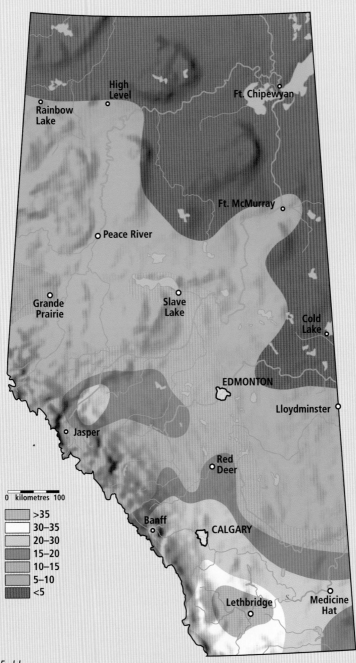

High Level
Ft. Chipewyan
Rainbow Lake
Ft. McMurray
Peace River
Grande Prairie
Slave Lake
Cold Lake
EDMONTON
Lloydminster
Jasper
Red Deer
Banff
CALGARY
Lethbridge
Medicine Hat

0   kilometres   100

>35
30–35
20–30
15–20
10–15
5–10
<5

*Fig. 5-11*

103

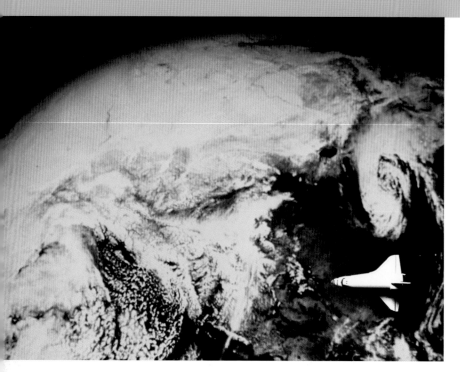

# Chapter Six: Optics and Acoustics

The atmosphere consists of a mixture of gases, which have optical and acoustic properties and can produce illusions as we observe and listen. To understand the optical effects, it is first necessary to appreciate that the electromagnetic radiation emitted by the sun extends over a wide range or spectrum of wavelengths. The solar spectrum starts with the shortest and most energetic waves, known as gamma and x-rays, and then extends into the ultraviolet wavelengths. Visible light is in the middle of the spectrum, and infrared radiation, which we sense as heat, has the longest wavelengths.

## Blue Sky

Why is the sky blue? This question was answered in the 19th century, when scientists realized that gas molecules scatter electromagnetic radiation. As sunlight enters the top of the atmosphere, the gases immediately start scattering some waves in the light spectrum. The size of the gas molecules determines which wavelengths are scattered. The atmosphere consists of mostly oxygen and nitrogen, and the molecules of these gases are the right size to scatter blue light while leaving the longer wavelengths basically untouched. The blue light rays bounce off other molecules as well. This scattering occurs in all directions—back toward the sun, at right angles and even at small angles to the solar rays—and gives the sky its apparent blue colour from the horizon to the sun. If one were up at the top of the atmosphere where there is little oxygen or nitrogen, the sky above would appear black—there is nothing to scatter light. As one progresses toward the earth's surface, the atmosphere takes on a dark blue colour as some blue light is scattered. Indeed, at the tops of mountains, the sky generally has a darker blue colour.

Alberta is known for its deep blue skies. There are two reasons for this: terrain in Alberta is relatively high, which results in naturally deeper blue skies, and the climate is also relatively dry and unpolluted (at least in the past). Water vapour molecules are large enough to scatter the all the visible wavelengths, so the presence of water vapour results in a whitish haze when looking toward the horizon. Clouds consist of large water droplets that scatter all wavelengths, making the clouds appear white. Atmospheric pollutants consist of gases, each of which has its own characteristic molecular makeup and size. These gases have an ability to scatter back incident light of certain wavelengths. The pollutants that are responsible for what is known as photochemical smog scatter light back into the yellow part of the spectrum. That's why the elements of smog appear to have a yellowish tint.

## Why the Sky is Blue

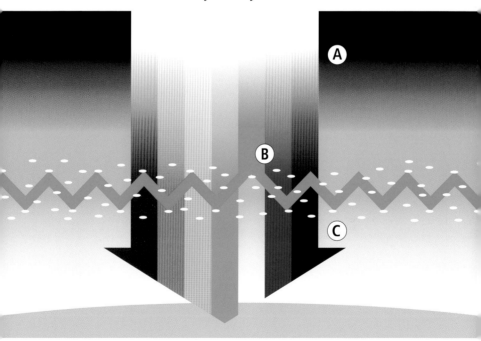

Fig. 6-1
A - white sunlight at the top of the atmosphere is composed of the full colour spectrum
B - some light in the blue colour spectrum is scattered in all directions by gas molecules in the atmosphere. As a result, we see blue sky in all directions.
C - the remaining light in other frequencies travels to the bottom of the atmosphere

These human-made, or anthropogenic, pollutants are more evident around the major cities in Alberta. In summer and winter downwind of both Edmonton and Calgary, there is a brownish to orange bank of pollution that originates from the city's vehicles and industrial emissions. Unfortunately, Alberta's characteristic blue skies are becoming something of a memory as industrialization increases.

# Rainbows

Falling water droplets characteristically have a spherical shape. When a light ray passes through a water droplet, it reflects off the back of the inside surface and then passes back out the front. Each time the ray moves between the air and water surface, some refraction occurs and, like the action of a prism, the white light spreads into its colour spectrum. A rainbow is the observed action of millions of droplets acting together. The angles between the sun, the raindrops and the observer are all well defined by optical theory with the result that the rainbow is essentially an arc that appears when the sun is at the observer's back, shining on a mass of raindrops. So, the rainbow appears in the opposite direction of the sun when the sun is relatively low in the sky and is illuminating a falling mass of raindrops. If we were to observe a bank of raindrops from an

Old weather saying:
*Rainbow in the morning gives you fair warning.*

## Formation of Primary and Secondary Rainbows

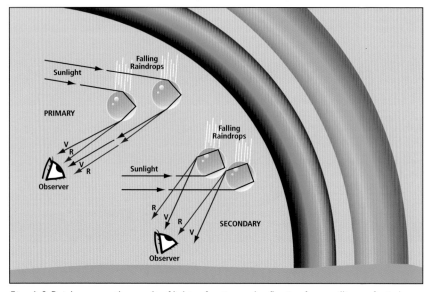

*Fig. 6-2* Rainbows are the result of light refracting and reflecting from millions of raindrops.

Primary and secondary rainbows

airplane or mountaintop, we would see a complete circle.

What has been described above is known as the primary rainbow—the result of one reflection off the back of the rain droplet. Because the droplets are spherical, in reality multiple reflections occur inside each drop. This gives rise to the appearance of a second or even third rainbow. With each reflection and refraction, there is some loss of brightness or intensity. As a result, the secondary bow is significantly less bright than the primary bow. This secondary bow is located outside the primary bow, and the colour

sequence is reversed. The tertiary bow, when observed, appears as an extremely faint arc inside the primary bow.

In theory, multiple bows are possible, but because of increasing loss of intensity, they are not observed.

*In the secondary rainbow, the colour scheme is reversed; instead of having red on the outside and blue on the inside, as do primary rainbows, the outside is blue and the inside is red.*

Sundogs form on both sides of the sun.

## Sundog Formation

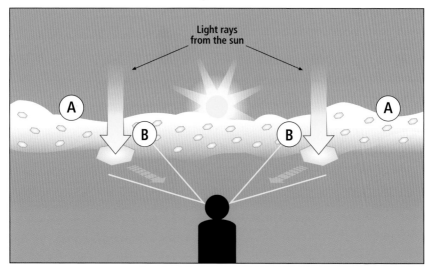

*Fig. 6-3*
A - cloud contains flat ice crystals
B - sun enters crystals and is bent at a 22 degree angle toward observer. Sundogs form
on both sides of sun

# Ice Crystal Phenomena

Several phenomena observed in Alberta's winter skies are a function of the unique characteristics of atmospheric ice crystals. The shape of the water molecule determines the shape of the resulting ice crystals. When formed in calm wind conditions, ice crystals are often six sided, or hexagonal. Sunlight can reflect off the ice crystal surfaces and can also pass through the crystal, which splits the white light in a prism-like fashion. This bending and splitting of light is known as refraction. Two commonly observed winter phenomena result: sundogs and light pillars.

**Sundogs** or, more correctly, parhelia are observed when the sun passes through a layer of uniform ice crystals. Refraction of the sun's rays at the air-ice surface of each ice crystal produces the characteristic prismatic colour band. When the ice crystals lay uniformly in a horizontal plane, the refracted light is concentrated enough to form the sundogs, which can be seen at an angle of 22 degrees on either side of the sun. On cold winter days, a layer of ice crystals may form near ground level. These ice crystals are often called diamond dust because they reflect sunlight, giving the impression of glittering diamonds. On cold, calm winter mornings, the combination of diamond dust and ice crystals aloft gives the most spectacular displays of sundogs.

## Light Pillar from Sunlight

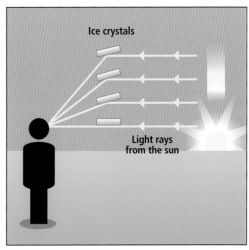

Fig. 6-4
Light reflected off underside of ice crystals results in a light pillar extending above the sun near the horizon

## Light Pillar from Street Light

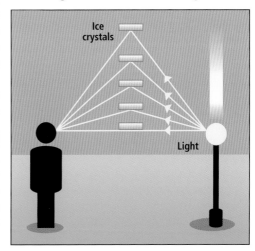

Fig. 6-5
Light reflected off ice crystals results in a light pillar extending above a street light

Light pillar extending above the sun

A **light pillar** is a reflection phenomenon that is formed in the presence of diamond dust. The more-or-less uniformly oriented ice crystals act individually as tiny reflecting surfaces. As shown in Figures 6-4 and 6-5, the resulting effect is a shaft of light that seems to extend above and below the light source. Any relatively intense light source will produce light pillars. During cold, calm nights, streetlights often appear to have pillars extending vertically above them.

## Mirages

A mirage is a refraction phenomenon that occurs when light passes between layers of air with different densities. Light rays are bent as they pass from colder air to intensely heated air close to the earth's surface. Such conditions happen when the sun heats sand or asphalt surfaces. Temperatures in the lowest metre or so above such surfaces can often vary up to 10° C, resulting in a significant air density difference between immediately adjacent layers of air. This bends the light rays passing between the air layers. The bending is not pronounced, but when we look toward the horizon, we

## The Mirage Effect

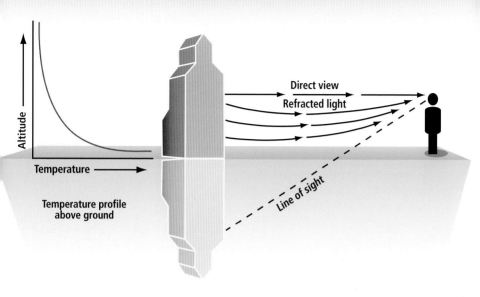

*Fig. 6-6* Objects on the horizon may appear inverted.

often see an intense blue image, which is in reality the bending of light from above the horizon. Occasionally it is not the sky that we see but rather the image of an object on the horizon. In this case, the image is turned upside down and is usually distorted. Mirages are generally unstable because the extreme vertical temperature differences also cause the warmer, lighter air parcels at ground level to move upward and be replaced by denser, cooler air immediately above. The result is turbulence in the mirage layer, resulting in an unstable, shimmering effect.

## A Mirage

An inverted image of a hilltop in a mirage

# Looming

This optical effect is similar to the mirage except that light rays are bent toward rather than away from the horizon. The condition happens when warm air is located above a colder layer near the earth's surface, which frequently occurs in the morning, when the air adjacent to the colder ground is overlain by warmer air aloft. Because this event occurs most mornings, looming is a relatively frequent occurrence, usually manifested by an apparent tilting up of the horizon. As a result, at sunrise it can seem to observers as though they are in a broad, shallow bowl. This phenomenon also affects shorter wavelengths, which affects the behaviour of weather radar beams. Radars emit microwave radiation, which behaves much like light waves do. Weather radars emit their microwave beam at a slight angle above the horizon. However, during strong inversion conditions, the radar rays are bent toward the ground, even beyond the true horizon. The ground reflects these rays back along the same path through the atmosphere to the radar receiver. As a result, the radar display shows an image of ground clutter for some distance around the radar's location. In effect, the radar is "seeing" the bottom of the shallow bowl.

# Other Optical Effects

Because much of weather observation depends on watching the sky, people throughout the centuries realized that they could interpret what the weather would be like in the short range by observing sky and cloud conditions. Much of this observation became folklore, but some of it has an element of scientific

rationale. An often-heard expression known to sailors is "red sky at night, sailor's delight; red sky at morning, sailor take warning." The explanation for this saying comes, in part, from the atmosphere's light-scattering capabilities, and from the direction of storm motion at mid-latitudes. At sunrise and sunset, the sun's rays are travelling through a long pathway, and the residual light is mostly red, giving cloud surfaces a reddish tint. At sunset, if there is no cloud in the western sky, clouds illuminated to the east look red. At mid-latitudes, storms move from west to east, and no cloud to the west indicates there are no approaching storms. So, "red sky at night [i.e., at sunset], sailor's delight" means no storms are on their way. If the sky is red in the morning, the rising sun is not obscured by cloud, and it may be illuminating the underside of a deck of cloud to the west. Again, because storms move from the west to east, it means potential storm clouds are approaching. Therefore, "red sky at morning [i.e., at sunrise], sailor take warning" means a storm is on its way. So, though it is folklore, the old saying has some basis in scientific reality.

# Acoustical Effects

Like light waves, sound waves propagate through the atmosphere, and they have some similar behavioural properties that are notable under inversion conditions. The most frequently observed effect is that of reflection at discontinuities in the atmosphere. Sound waves are bent or refracted at relatively sharp discontinuities in the vertical thermal structure. At night, nocturnal temperature inversions form when air in contact with the earth's surface cools more

## Refraction of Sound Waves

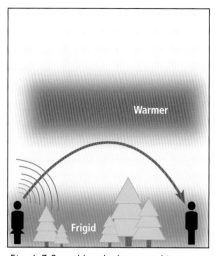

*Fig. 6-7* Sound bends downward in a temperature inversion

Sound bends upward if surface is warm

quickly than the air aloft, resulting in the reverse of the normal situation in which temperature drops off slowly with height. In this inversion situation, there can be a sharp discontinuity in temperature. Sound waves propagating at an angle toward this discontinuity are refracted downward toward the earth's surface. This phenomenon is most readily apparent when the observer is relatively distant from a steady source of noise, such as roadway with heavy traffic. At night, the noise from a distant highway, train or even general traffic noise from a city seems to be louder than it is during the day. Of course, other factors such as background noise levels and wind direction also come into play, but the effect can be quite noticeable, for example when one normally does not hear a freeway, but it becomes quite notable after sunset when the inversion becomes well established. Sound waves can also be bent upwards when temperatures decrease rapidly with

height. For example, we often do not hear thunder from a distant lightning stroke because the sound wave is bent upward and away from us.

A simple acoustic effect relates to the rate of propagation of sound waves in air. Sound waves propagate at a speed of approximately 330 metres per second in the lower atmosphere. Light, on the other hand, is transmitted essentially instantaneously. We can use this difference to give a rough estimate of the distance to a lightning stroke by counting the number of seconds between seeing a lightning flash and hearing the crack of thunder. A pause of three seconds means the sound travelled about 1 kilometre, six seconds indicates 2 kilometres and so on. The timing of distant lightning strokes becomes more complicated because of reflection of the sound off atmospheric discontinuities, such as edges of clouds and shafts of falling rain.

# Chapter Seven: Alberta's Climate

## Controls and Influences

The climate of a region is often described by statistics, such as averages, extremes and the most and least frequent occurrence of specific weather elements (precipitation, temperature, etc.). In reality, climatic averages are only lines on a graph. The day-to-day weather varies around these averages, from one weather condition to another. The climate defines what we can do comfortably, when we can do it, and what conditions may prevent certain activities. As the old saying goes, climate is what you may expect but weather is what you get. Overall, Alberta's climate can be classified as continental because the province is relatively far removed from large water bodies. It features a wide variation in temperature and precipitation from summer to winter.

Alberta is subjected to an assortment of air masses, as illustrated in Figure 7-1. These air masses represent the characteristics of the regions in which they were formed. In winter, air masses that originate over arctic regions of Canada or Asia flow over the northern Pacific Ocean, taking on its heat and moisture. This maritime arctic air is a frequent visitor to Alberta. Maritime polar air originates over the mid-latitude Pacific Ocean and is predominant in summer. In winter when the atmospheric circulation sets up from the southwest, Alberta is subjected to what is called the "pineapple express," a warm, moist maritime polar air mass that originates in the

## Winter Air Masses and Circulation

**Continental Arctic**
- very cold, -25 to -50° C
- dry, very stable
- pronounced temperature inversion

**Maritime Arctic**
- very unstable
- clouds, frequent showers or flurries
- visibility good except in showers

**Maritime Polar**
- milder and more stable than arctic air

**Pacific Maritime Tropical**
- light winds, cooler than Atlantic air
- comes to North America from west or northwest
- stable in lower 1000 m (marine stratum)

**Atlantic Maritime Tropical**
- comes to North America from south or southeast
- warm and humid

**SST** - Sea surface temperature

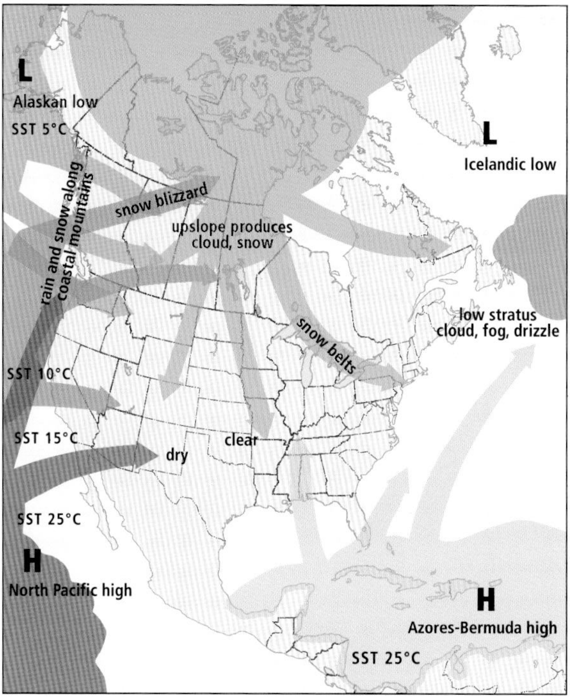


## Summer Air Masses and Circulation

**Continental Tropical**
- hot, dry, unstable

**Maritime Arctic**
- continental air modified by open seas, lakes and swamps

**Maritime Polar**
- warmer and more stable than maritime arctic air

**Pacific Maritime Tropical**
- high pressure blocks moist air

**Atlantic Maritime Tropical**
- oppressively hot and humid
- unstable, frequent thunderstorms

**SST** - Sea surface temperature

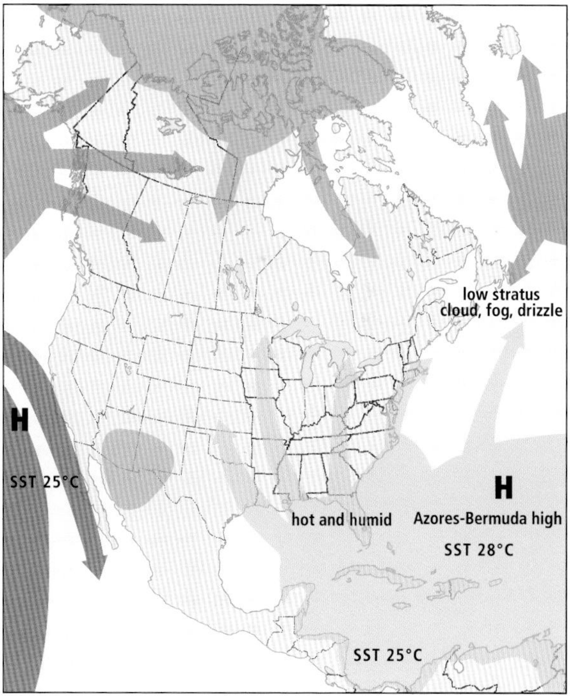


*Fig. 7-1*

tropics. The Rocky Mountains modify these Pacific air masses, wringing out much of the moisture as the air climbs over high terrain and pushes eastward across the province. Continental arctic air from northern Canada and over the frozen Arctic Ocean is also common in Alberta. This air mass is barely modified as it moves southward, unimpeded by mountain ranges—it is cold and extremely dry. On rare occasions, warm, moist air formed over the tropical Pacific Ocean, the Caribbean Sea and Gulf of Mexico moves as far north as central Alberta. When this air arrives, it usually spells trouble, resulting in periods of heavy rain and the most severe summer thunderstorms. Alberta is rarely considered a source region for air masses, but it can, on occasion, be part of the continental polar air mass source region of the southwest United States. Prolonged hot, dry conditions occur over the southern half of the province when such an air mass predominates.

The global circulation pattern is one major influence on Alberta's climate. Westerly winds in the upper atmosphere guide the motion of weather systems over the Alberta landscape. These weather systems provide much of the day-to-day variation in weather conditions. Frontal zones, which are the boundaries between the air masses, are part of these systems. Light rainfalls or showery conditions in the warm season are usually associated with frontal weather systems. Storms associated with fronts are stronger in winter than in summer and can bring about blizzard conditions.

The global atmospheric circulation pattern undergoes periodic changes, and El Niño of the tropical Pacific Ocean has a significant influence on the climate in Alberta as well as in other parts of the world. This phenomenon demonstrates the link between circulation patterns in the atmosphere and those in the earth's oceans. The main currents of the Pacific basin are shown in Figure 7-2. The trade winds that blow from east to west along the equatorial Pacific drag surface ocean waters toward the west. Accompanying the westward drift of water is an upwelling of cool, deep, nutrient-rich water along the South American coast, which gives rise to a productive fishing industry off Peru and Ecuador. The trade winds occasionally slacken or even reverse, and the accumulated water off the Indonesian coast begins to shift back across the Pacific, taking about two months to reach the coast of South America. When the bulge of warm water reaches the coast of South America, it cuts off the upwelling of the nutrient-rich waters. The warming event usually starts in December, and

| Extreme Windchills—Alberta Stations | | |
|---|---|---|
| Location | Maximum (° C) | Date |
| Banff | -52.1 | Jan. 16, 1970 |
| Calgary | -55.1 | Dec. 15, 1954 |
| Edmonton | -61.1 | Jan. 26, 1972 |
| Fort McMurray | -59.6 | Feb. 27, 1962 |
| Grande Prairie | -63.0 | Jan. 14, 1953 |
| High Level | -57.1 | Jan. 25, 1972 |
| Jasper | -53.6 | Jan. 25, 1972 |
| Lethbridge | -55.7 | Dec. 28, 1968 |
| Lloydminster | -54.5 | Feb. 7, 1994 |
| Medicine Hat | -58.9 | Dec. 28, 1968 |
| Peace River | -58.4 | Dec. 30, 1992 |
| Red Deer | -60.1 | Jan. 26, 1972 |
| Fort Chipewyan | -59.1 | Dec. 11, 1975 |
| Cold Lake | -55.4 | Feb. 27, 1962 |
| Slave Lake | -56.8 | Dec. 30, 1992 |

## Pacific Ocean Currents

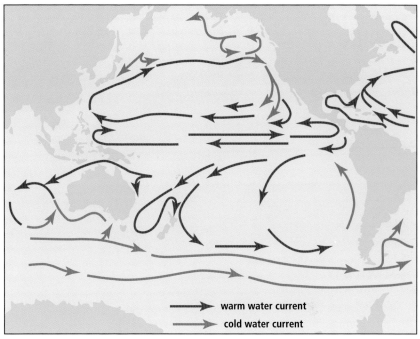

warm water current
cold water current

*Fig. 7-2*

the local fisherman named it El Niño, Spanish for "the little boy" and a reference to the Christ child. When the waters in the eastern Pacific are colder than normal, what is known as La Niña (or "the little girl") occurs. La Niña is the antithesis of El Niño.

El Niño warming events occur on an irregular basis at intervals of two to seven years. Accompanying this oceanic cycle is an oscillation in atmospheric pressure on either side of the Pacific Ocean. At the onset of El Niño, average pressures over the western Pacific are higher than normal and are accompanied by lower than normal pressures in the east. This seesaw cycle is known as the Southern Oscillation. The ocean and atmospheric cycles

are linked and are known as the ENSO or El Niño-Southern Oscillation.

Strong El Niño and La Niña episodes have a noticeable effect on the weather conditions in western North America. During an El Niño event, a stronger area of low pressure prevails over the west coast of North America, and the jet stream across the northern Pacific takes a southerly track. This has the effect of directing warmer than normal air into Alberta. Therefore, during a strong El Niño event, winters in Alberta are milder than average. However, during a strong La Niña, the normal area of low pressure off the BC coast is replaced by higher pressures. A split in the jet stream pattern accompanies the pressure change, and

## Typical January–March Weather Anomalies and Atmospheric Circulation During Moderate to Strong El Niño & La Niña

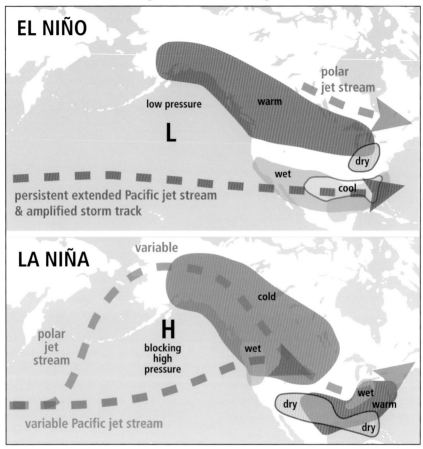

*Fig. 7-3*

the northern portion of the jet stream moves over Alaska before heading southeastward along the Continental Divide. Storms and cold air masses are guided by the jet stream, so winters are colder than average during a La Niña episode. The locations of the jet streams during these events are shown in Figure 7-3.

During El Niño winters, Alberta experiences below-average snow cover, and temperatures are well above normal. One of the strongest El Niño events occurred in the winter of 1982–83 when there were widespread impacts across the globe. Central Alberta received less than 60 percent of its usual amount of winter precipitation, and temperatures were above normal. An El Niño of moderate strength occurred in 2002–03. December of 2002 was largely snow-free across western Canada, and temperatures were well above normal across Alberta for most of the winter. Calgary had an average December temperature that was nearly 8° C above normal.

La Niña events are most pronounced during winter months, affecting temperatures and precipitation. La Niña of 1988–89 saw winter temperatures below normal over southern Alberta, and several heavy snow and blizzard events occurred across the province during the winter of 1989. On January 30, 1989, Edmonton received 33 centimetres of snow, setting a new record in the city for the most snowfall in one day in January. A moderate La Niña in the winter of 2007–08 brought colder than normal conditions across the northern half of Alberta. This La Niña will be remembered in eastern Canada as a winter of abnormally high snowfall.

The climate across the province is also determined by what is known as the physiography of the landscape. The Rocky Mountains are the most pronounced feature, and they significantly modify the character of air masses and the behaviour of weather systems. Low pressure centres often form in the lee of the Rockies and are sometimes accompanied by strong easterly winds, which ascend the lee slopes of the mountains. In such conditions, heavy rainfalls over the headwaters of river systems can result in flood conditions, especially during the spring runoff period.

High terrain modifies temperatures, especially at night in the foothill areas of western Alberta. Locations in the foothills experience cooler overnight temperatures than more northern locations. Ranges of higher hills can also affect the climatic

character. The Cypress Hills of extreme southeast Alberta and the Swan Hills of central Alberta are areas where thunderstorms often form because of the combined effects of low level moisture and the intense solar heating on south facing slopes. River valleys also influence heating and cooling. At night, colder air may pool along river valleys, resulting in earlier frosts than at nearby locations outside of the valleys. South facing slopes are popular with farmers because the slopes get more direct sun and are therefore warmer, especially early in the growing season.

The northern half of the province is dominated by the boreal plain, which consists of a mixture of forested areas dotted with lakes and a few relatively significant areas of glacial remains, including the Swan Hills, the Birch Mountains and the Caribou Mountains. These topographic features can produce their own mini climate zones. The prairie grassland region, a relatively flat, open area characterized by grasslands, dominates the southern half of the province.

## Temperature

Temperature is the weather element that has the greatest influence on living entities. Temperature variations largely control the viability of plants, and temperature means and extremes define human comfort levels. In Alberta, the temperature pattern across the province is an indication of the variation of the sun's angle from the south to the north. Annual mean temperatures are warmest in the region from Lethbridge to Medicine Hat and drop off at a fairly uniform rate as we move toward the northern border.

The temperature variation through the year is a reflection of Alberta's continental climate. Winters are long and cold, whereas summers are short and relatively cool. The spring and fall transition, or shoulder seasons, are short and often include stretches of weather more like that of the primary seasons. Albertans are accustomed to having a heavy snowfall in late spring and early autumn.

Alberta's summer temperatures show the strong terrain influence across the province. The July daily mean temperature pattern in Figure 7-5 shows a pronounced east/west gradient much like the elevation gradient. There is relatively little variation in average July temperatures from south to north. Although the warmest summer temperatures are in the lower elevation areas of southeastern Alberta, the Fort Chipewyan area at the northern border has July average temperatures that are about the same as Edmonton and Lethbridge. Summer temperatures only occasionally attain extreme values. The

## Apparent Temperature

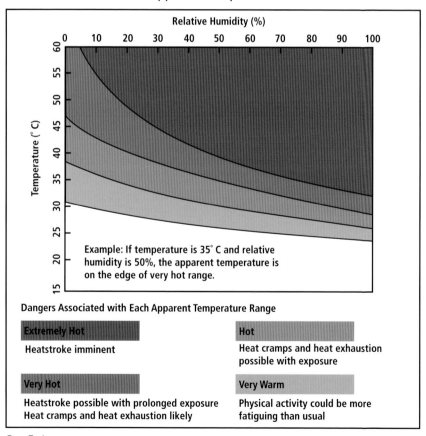

Example: If temperature is 35° C and relative humidity is 50%, the apparent temperature is on the edge of very hot range.

**Dangers Associated with Each Apparent Temperature Range**

**Extremely Hot**
Heatstroke imminent

**Hot**
Heat cramps and heat exhaustion possible with exposure

**Very Hot**
Heatstroke possible with prolonged exposure
Heat cramps and heat exhaustion likely

**Very Warm**
Physical activity could be more fatiguing than usual

*Fig. 7-4*

## July Daily Mean Temperature

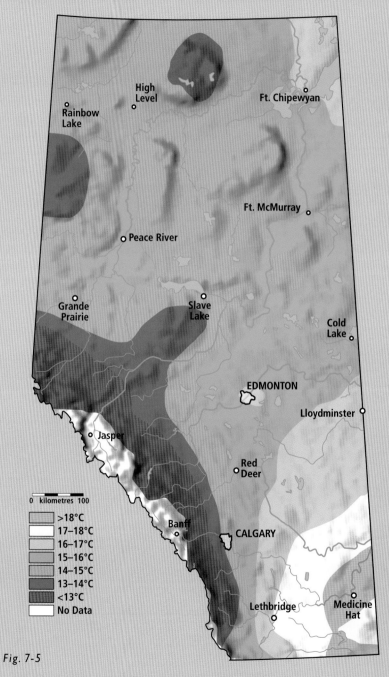

Rainbow Lake

High Level

Ft. Chipewyan

Ft. McMurray

Peace River

Grande Prairie

Slave Lake

Cold Lake

EDMONTON

Lloydminster

Jasper

Red Deer

Banff

CALGARY

Lethbridge

Medicine Hat

0    kilometres    100

| >18°C |
| 17–18°C |
| 16–17°C |
| 15–16°C |
| 14–15°C |
| 13–14°C |
| <13°C |
| No Data |

*Fig. 7-5*

highest temperature recorded was 43.3° C at Fort Macleod on July 18, 1941. Alberta's hottest summers occur in the southeast, and Medicine Hat—where the average high temperature for June, July and August is 25.8° C—has the fifth hottest summer in Canada.

Winter temperatures show a pronounced gradient from south to north, as shown in Figure 7-7. This again indicates the importance of sun's angle in winter and the tendency for cold outbreaks of arctic continental air masses to push southeastward across the province. The influence of terrain is also evident; mean temperatures in the Lethbridge-to-Calgary corridor average higher than - 10° C, which is a reflection of the frequency of Chinooks. The Chinook can provide a welcome respite from winter in southwestern Alberta. Typically a Chinook is accompanied by rapid warming

and strong, gusty winds. An extreme example occurred at Pincher Creek in January 1962. As shown in Figure 7-6, during this event, the temperature rose from -20° to 3° C in one hour; winds went from being nearly calm to gusting from the west-southwest at speeds of up to 110 kilometres per hour.

A practical application of temperature statistics is the calculation of the Heating Degree Day (HDD). Heating Degree Days (see Figure 7-8) are highly correlated with the amount of energy required to heat buildings to a comfortable level. Heating is generally required when external temperatures are below some reference value, usually 18° C. The Heating Degree Day is calculated by subtracting the average daily temperature from 18. For example, if the average temperature for a day is 10° C, then the HDD value would be 8; or if the day's average is

## Temperature & Wind Trace—Pincher Creek January 27, 1962

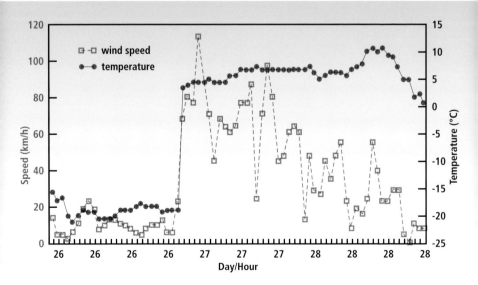

*Fig. 7-6*

# January Daily Mean Temperature

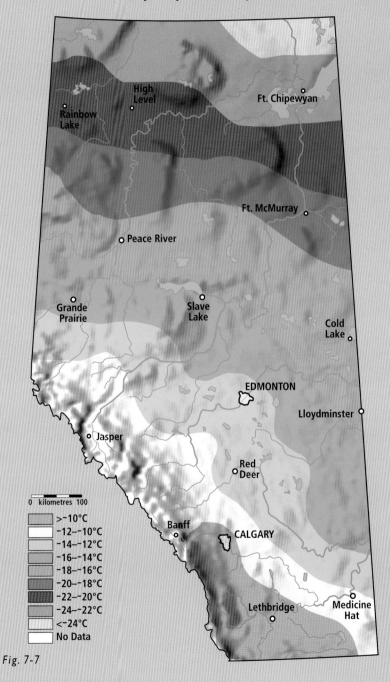

0  kilometres  100

- >-10°C
- -12--10°C
- -14--12°C
- -16--14°C
- -18--16°C
- -20--18°C
- -22--20°C
- -24--22°C
- <-24°C
- No Data

Fig. 7-7

- 10° C, then the HDD value is 28. Daily temperatures above 18 have an HDD of zero. The HDD values can be summed over a year to give a useful estimate of heating energy requirements. As shown in Figure 7-9, Heating Degree Days

## What Are Heating Degree Days?

range from their lowest values of near 4600 in the Medicine Hat-to-Lethbridge corridor to more than 7500 along the northern border. This means that it would take approximately twice the amount of energy to heat a home in Fort Chipewyan as it would in Medicine Hat. Edmonton and Calgary have about the same Heating Degree Day values—about 5200.

The summer growing season varies greatly across Alberta. The Growing Degree Day (GDD) is a useful measure of the amount of heat available to initiate and sustain crop growth. The GDD is related to how far the average daily temperature varies from a specific reference temperature. The daily values are accumulated over the growing season. The reference temperature depends on the type of plant, but 5° C is often used for general plant growth. Figure 7-10 shows the Growing Degree Days above 5° C for Alberta. The area between Medicine Hat and Lethbridge has the most Growing Degree Days, and the Peace River region around Grand Prairie also has a high GDD. Plant growth, of course, depends on the frost-free period as well. The frost-free period is defined as the number of days between the last frost in spring and the first frost in autumn. Figure 7-11 shows the average frost free period across the province. Much of the grassland area of southeastern Alberta has more than 125 frost-free days per year.

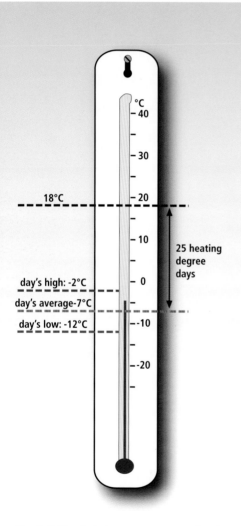

*Fig. 7-8* The day's average temperature is subtracted from 18° C to get the heating degree days.

## Annual Total Degree Days Below 18° C

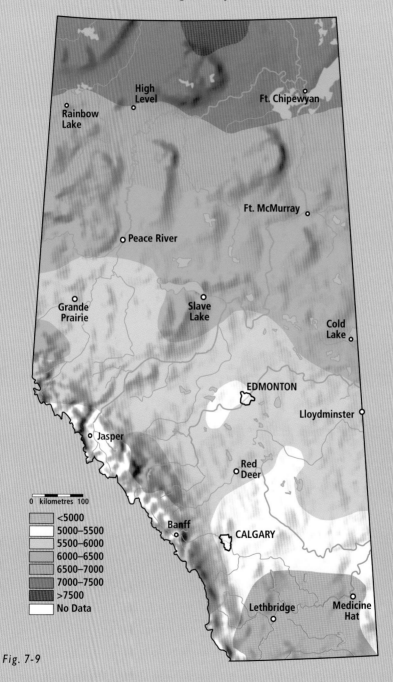

*Fig. 7-9*

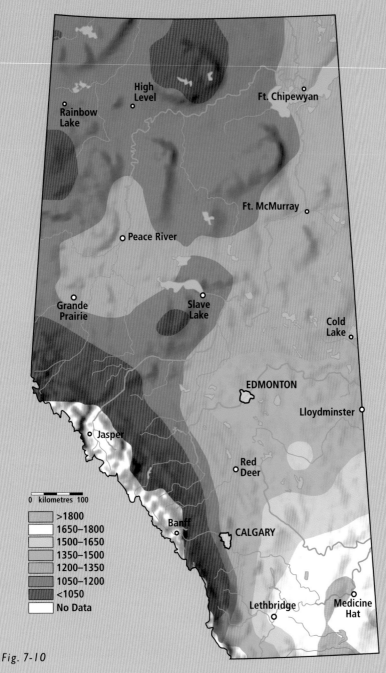

## Annual Total Degree Days Above 5° C

High Level
Rainbow Lake
Ft. Chipewyan
Ft. McMurray
Peace River
Grande Prairie
Slave Lake
Cold Lake
EDMONTON
Lloydminster
Jasper
Red Deer
Banff
CALGARY
Lethbridge
Medicine Hat

0    kilometres   100

| | |
|---|---|
| | >1800 |
| | 1650–1800 |
| | 1500–1650 |
| | 1350–1500 |
| | 1200–1350 |
| | 1050–1200 |
| | <1050 |
| | No Data |

*Fig. 7-10*

## Frost-free Period: Days Above 0° C

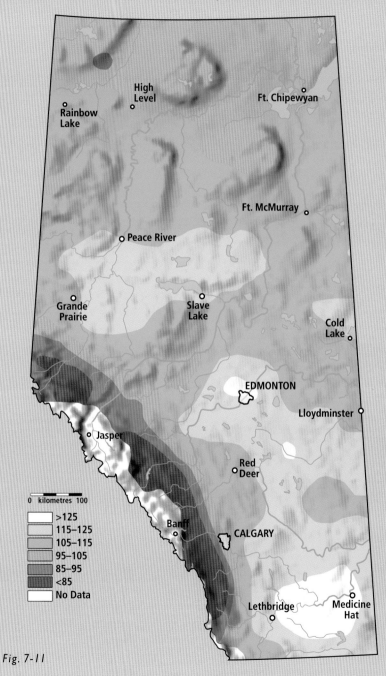

High
Level

Ft. Chipewyan

Rainbow
Lake

Ft. McMurray

Peace River

Grande
Prairie

Slave
Lake

Cold
Lake

EDMONTON

Lloydminster

Jasper

Red
Deer

0    kilometres  100

Banff

CALGARY

>125
115–125
105–115
95–105
85–95
<85
No Data

Lethbridge

Medicine
Hat

*Fig. 7-11*

Cities have a modifying effect and create local microclimates, known as urban heat islands. The most apparent effect is on temperature because buildings, factories and moving vehicles emit large amounts of waste heat, whereas concrete buildings, asphalt parking lots and roadways absorb heat during daylight hours and release it overnight. The larger the urban area, the more pronounced the heat island. The heat island effect in Alberta's two largest cities amounts to several degrees Celsius and is most pronounced in winter, especially with minimum overnight temperatures. This is perhaps best demonstrated by comparing the temperature records at the Edmonton International and Edmonton City Centre airports. The international airport is located some 15 kilometres south of the edge of the urban buildup, and both airports are at about the same elevation above sea level. Average minimum temperatures in January and average maximums in July are about 3° C warmer at the City Centre Airport. Differences can be even more pronounced on calm nights in winter, when overnight minimums can differ by nearly 10° C. The urban heat island effect is also evident in the calculation of the frost-free period. Edmonton City Centre Airport has the longest average frost-free period in Alberta at 143 days, whereas the average frost-free period at the Edmonton International Airport is 115 days.

*In Alberta, 30 to 35 percent of the precipitation that falls is in the form of snow.*

# Moisture and Precipitation

Precipitation across Alberta is not uniform throughout the year. Snowfall accounts for only 30 to 35 percent of the total precipitation across the province. The majority of the precipitation falls during summer, when it is most needed for agriculture. Summer precipitation usually comes in the form of showers, and it is a rare day when there is not a rain shower occurring somewhere across the province. Figure 7-12 compares the monthly precipitation at three stations, and the early-summer peak is clearly evident.

Total precipitation across Alberta is shown in Figure 7-13. Annual precipitation ranges from less than 300 millimetres per year in the southeast to more than 600 millimetres per year in the foothills of central Alberta. Southeastern Alberta is part of a region known as Palliser's Triangle, named after John Palliser, who surveyed the area in the late 1850s. He believed this region of low precipitation and warm summer temperatures would be unsuitable for growing crops.

Rainfall can occasionally come in the form of a prolonged drenching that is usually associated with what is known as a cold low. This type of circulation system moves into Alberta during early summer and can remain almost stationary for several days in the vicinity of the Rocky Mountains. Winds flowing counter-clockwise around a cold low often contain an abundance of moisture, and rainfall is persistent. The highest recorded one-day rainfall—213 millimetres in Eckville—was associated with such a cold low circulation. This type of storm can soak the surface soil layer, leading to flooding in the river basins of the western half of the province.

Alberta is not known as a region of heavy snowfall. The average annual snowfall is shown in Figure 7-14 and is about 150 centimetres per year across the province. The southeast area from north of Medicine Hat to Drumheller and Oyen records just less than 100 centimetres of snow per year. The foothills receive about 200 centimetres per year, whereas the mountains can get more than 400 centimetres annually. As shown on the map, snow is on the ground for an average of 150 to more than 170 days in the northern half of the province and for about 110 to 120 days in the southern grassland regions. But the variation from year to year is extreme. In the winter of 2005–06, most of south and central Alberta did not have a continuous snow cover until the third week in February. The following year, snow was on the ground in central Alberta by the end of October. In both years, the snow disappeared by mid-April.

Snow can fall in any month of the year in Alberta, and most Alberta locations have recorded at least a trace of snow in every month. Late-spring storms with heavy snowfall can occur in any region of the province, and they are usually most unwelcome. When they take place in ranching country during the calving season, they can be particularly devastating. Access to herds can be limited, and forage grasses can be covered with a thick layer of wet snow.

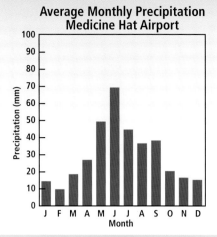
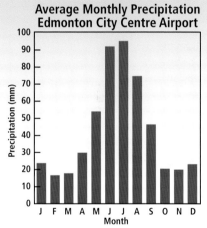
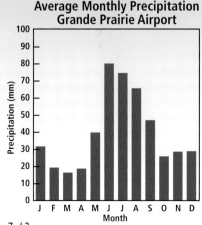

Fig. 7-12

129

# Average Annual Precipitation

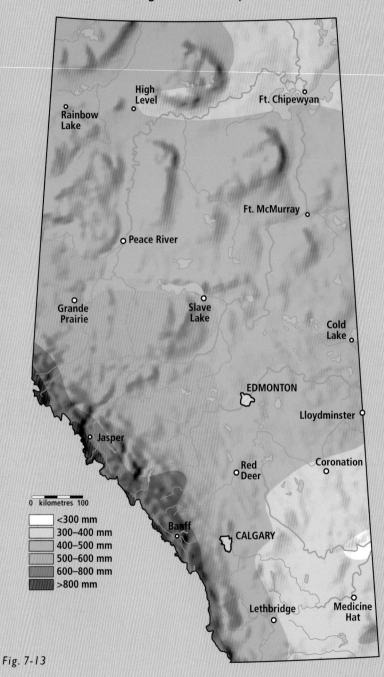

High
Level

Ft. Chipewyan

Rainbow
Lake

Ft. McMurray

Peace River

Grande
Prairie

Slave
Lake

Cold
Lake

EDMONTON

Lloydminster

Jasper

Red
Deer

Coronation

Banff

CALGARY

Lethbridge

Medicine
Hat

0   kilometres 100

| | |
|---|---|
| | <300 mm |
| | 300–400 mm |
| | 400–500 mm |
| | 500–600 mm |
| | 600–800 mm |
| | >800 mm |

*Fig. 7-13*

# Average Annual Snowfall

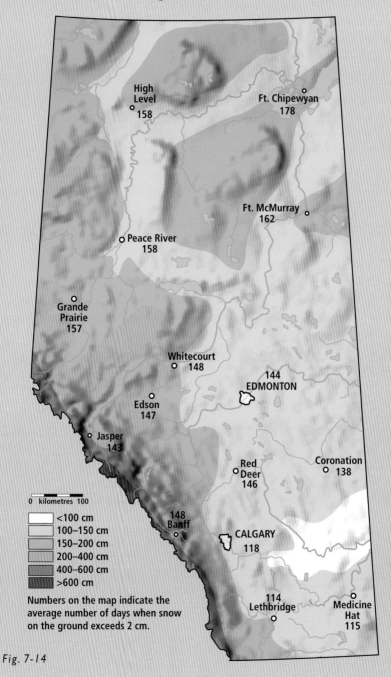

High Level 158

Ft. Chipewyan 178

Ft. McMurray 162

Peace River 158

Grande Prairie 157

Whitecourt 148

144 EDMONTON

Edson 147

Jasper 143

Red Deer 146

Coronation 138

0   kilometres 100

<100 cm
100–150 cm
150–200 cm
200–400 cm
400–600 cm
>600 cm

148 Banff

CALGARY 118

Numbers on the map indicate the average number of days when snow on the ground exceeds 2 cm.

114 Lethbridge

Medicine Hat 115

Fig. 7-14

131

# Drought

The absence of precipitation is particularly problematic, and it affects all sectors of the economy. Reduced crop yields and crop failure are the most obvious impacts in the agricultural sector, but soil loss through wind erosion, reduction in municipal water supply, the increased susceptibility of timber to fire and pest infestation, and the loss of wildlife habitat are also serious concerns. Drought conditions occur because of a number of variables, including the lack of precipitation, the variability of precipitation with time

(i.e., precipitation at the wrong time), concurrent warm spells and even changes in the ability of society and wildlife to adapt to a changing environment. The Palmer Drought Index (PDI) is used extensively in western Canada to depict drought conditions. Figure 7-15 shows the PDI calculated over the province as a whole and over agricultural areas from 1961 to 1991. Index values below zero represent dry conditions, values between -2 and -4 indicate moderate drought, and PDI values lower than -4 indicate extreme drought. When the PDI is calculated over

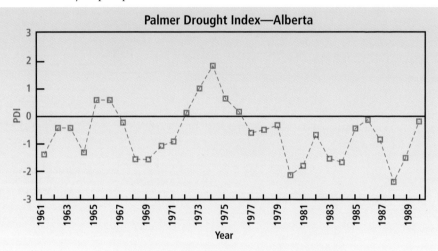

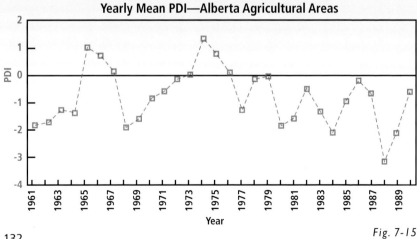

Fig. 7-15

Alberta, the long-term record is cyclical. Conditions were generally below the zero or normal line for most of the 30-year period, punctuated by a few wet periods about every five years.

# Combinations of Weather Elements

Human discomfort is usually caused by a combination of weather elements. In winter, the combination of cold temperatures and wind has a pronounced chilling effect on skin. In summer, high humidity combined with warm temperatures prevents our bodies' evaporative cooling mechanism from performing effectively.

## Wind Chill

The wind chill index has been developed to measure this cooling effect and to signal conditions that may be harmful. The wind chill calculation table provides "feels like" values in degrees Celsius, indicating how cold the temperature feels once the strength of the wind has been factored in. Wind chill values can fall below -30° C at all locations in Alberta during winter.

## Wind Chill Calculation Table

| T air / $V_{10}$ | 5 | 0 | -5 | -10 | -15 | -20 | -25 | -30 | -35 | -40 | -45 | -50 |
|---|---|---|---|---|---|---|---|---|---|---|---|---|
| 5 | 4 | -2 | -7 | -13 | -19 | -24 | -30 | -36 | -41 | -47 | -53 | -58 |
| 10 | 3 | -3 | -9 | -15 | -21 | -27 | -33 | -39 | -45 | -51 | -57 | -63 |
| 15 | 2 | -4 | -11 | -17 | -23 | -29 | -35 | -41 | -48 | -54 | -60 | -66 |
| 20 | 1 | -5 | -12 | -18 | -24 | -30 | -37 | -43 | -49 | -56 | -62 | -68 |
| 25 | 1 | -6 | -12 | -19 | -25 | -32 | -38 | -44 | -51 | -57 | -64 | -70 |
| 30 | 0 | -6 | -13 | -20 | -26 | -33 | -39 | -46 | -52 | -59 | -65 | -72 |
| 35 | 0 | -7 | -14 | -20 | -27 | -33 | -40 | -47 | -53 | -60 | -66 | -73 |
| 40 | -1 | -7 | -14 | -21 | -27 | -34 | -41 | -48 | -54 | -61 | -68 | -74 |
| 45 | -1 | -8 | -15 | -21 | -28 | -35 | -42 | -48 | -55 | -62 | -69 | -75 |
| 50 | -1 | -8 | -15 | -22 | -29 | -35 | -42 | -49 | -56 | -63 | -69 | -76 |
| 55 | -2 | -8 | -15 | -22 | -29 | -36 | -43 | -50 | -57 | -63 | -70 | -77 |
| 60 | -2 | -9 | -16 | -23 | -30 | -36 | -43 | -50 | -57 | -64 | -71 | -78 |
| 65 | -2 | -9 | -16 | -23 | -30 | -37 | -44 | -51 | -58 | -65 | -72 | -79 |
| 70 | -2 | -9 | -16 | -23 | -30 | -37 | -44 | -51 | -58 | -65 | -72 | -80 |
| 75 | -3 | -10 | -17 | -24 | -31 | -38 | -45 | -52 | -59 | -66 | -73 | -80 |
| 80 | -3 | -10 | -17 | -24 | -31 | -38 | -45 | -52 | -60 | -67 | -74 | -81 |

T air = air temperature in ° C and $V_{10}$ = observed wind speed at 10 m elevation, in km/h.

## Frostbite Guide

| |
|---|
| Low risk of frostbite for most people |
| Increasing risk of frostbite for most people in 10 to 30 minutes of exposure |
| High risk for most people in 5 to 10 minutes of exposure |
| High risk for most people in 2 to 5 minutes of exposure |
| High risk for most people in 2 minutes of exposure or less |

## Humidex

The degree of discomfort humans experience in warm weather depends on the rate at which our bodies can lose heat through the evaporation of perspiration. This natural cooling mechanism is compromised when atmospheric humidity is high enough to inhibit evaporation. The humidex index (Figure 7-16) combines air temperature and relative humidity into a number that is a perceived temperature. When humidex values are higher than 30, some people experience discomfort, and when the value is over 40, everyone is uncomfortable. Because Alberta is a dry province,

Alberta humidex values rarely exceed 30, averaging less than nine days per year.

## Sunshine

Alberta is perhaps best known for its clear blue skies and ample sunshine. Maritime air masses that frequent the province lose much of their moisture as they cross the Rocky Mountains, and the arctic air mass of winter is generally very dry, so Alberta usually has less cloud cover than the rest of Canada. Figure 7-17 shows the average hours of sunshine across the province, and as expected, southeast Alberta gets the most hours. Medicine Hat has the highest number of hours of sunshine in Canada. In fact, Medicine Hat and Calgary rank one and two in Canada as the locations having the most days with sunshine. Hours of sunshine drop off in the foothills, showing the effects of increased cloud formed in upslope flow conditions in that area.

| HUMIDEX (°C) | DEGREE OF DISCOMFORT |
|---|---|
| 20 – 29 | Comfortable |
| 30 – 39 | Varying degrees of discomfort |
| 40 – 45 | Almost everyone uncomfortable |
| 46+ | Active physical exertion must be avoided |
| 54+ | Heat stroke may be imminent |

*Fig. 7-16*

## Average Total Hours of Sunshine

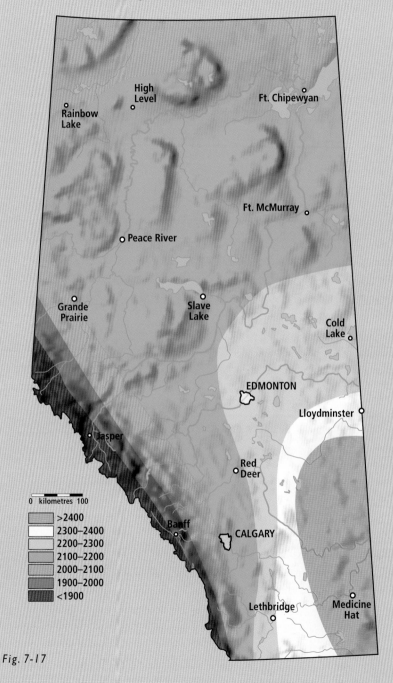

Fig. 7-17

135

# Chapter Eight: Storms

Albertans are subjected to almost all categories of severe weather. We suffer through tornadoes, hailstorms, monsoon-like rainfalls, windstorms and associated dust storms, blinding blizzards and monstrous snowfalls. We even get the occasional bout of freezing rain. At least in Alberta we are not subjected to tropical hurricanes.

Storms are a frequent topic of conversation for Albertans. Talk may be stimulated by significant weather events or a harrowing escape from the whims of Mother Nature. Often it is a recollection of a blizzard or tornado that resulted in loss of life. Sometimes it is the anticipation of tomorrow's forecast that is the concern. How much snow will I need to shovel? What are the chances an afternoon thunderstorm will catch me on the golf course, at my picnic or at the lake? In Alberta, storms are not predominant in any one season. Thunderstorms and their associated phenomena are the most common events of the warm season, and blizzards are an occasional event of the winter. Autumn seems to be the one season that is reasonably benign—but perhaps that is because it is a short season in Alberta. Occasionally, Mother Nature seems to get storms and seasons confused. A thunderstorm in winter is an infrequent visitor, whereas a late spring snowstorm and the havoc it can create seems to be an all too frequent event just when we are anticipating the beginning of summer.

Air masses interact along frontal zones, and the real action happens when a front is moving, strengthening or causing the explosive strengthening of storms. Alberta is located in the middle of this active zone of frontal jostling, so storms in this province are almost always associated with fronts.

## Winter Storms

### Blizzards

In Alberta, probably the most dramatic winter storms are what we commonly refer to as blizzards. To meet the official definition of a blizzard, a storm must have temperatures lower than 0° C, winds stronger than 40 kilometres per hour and visibility of less than 1 kilometre because of falling and blowing snow, and it must last longer than four hours. The amount of snowfall is not included in the definition—blowing snow is the issue. But even if severe winter storms do not meet the criteria of the official definition, they are not to be taken lightly. Blizzards do not only cause hardship, they can also cause loss of life. Environment Canada's website warns that, "Winter storms and excessive cold claim more than 100 lives every year in Canada, more than the combined toll from hurricanes, tornadoes, flood, extreme heat and lightning."

> Why, what's the matter, that you have such a February face,
>
> So full of frost, of storm, and cloudiness.
>
> —Shakespeare, *Much Ado About Nothing*

The blizzard of January 1973 was memorable for many Albertans. On January 1 and 2, temperatures across southern and central Alberta were well above zero, and they were around zero in

Blizzard conditions (above); a long roll cloud heralds the arrival of a cold front in early winter (opposite).

the Peace Country of northwestern Alberta. The situation was changing rapidly, however, and much colder air poised in northern Alberta started moving southward across the province on January 1. Overriding moist Pacific air released copious amounts of snow as the storm intensified and the circulation strengthened. Just after midnight on January 1, the cold front went through Edmonton, and the snow and blowing snow began. In the early morning hours of January 2, a freight-carrying jet aircraft crashed at Edmonton International Airport, killing all members of the crew. Throughout January 2 and most of January 3, blizzard conditions raged, with

wind chill temperatures near -40° C over most of central Alberta. I was a junior weather forecaster at the time and was returning to Edmonton on the evening of January 2. Blizzard conditions reduced visibility to near zero north of Calgary, and I had to wait out the storm in the town of Bowden, just south of Red Deer.

Another memorable January blizzard occurred at the end of January 1989. On January 30, temperatures reached highs of 11° C in Calgary and 6° C in Edmonton. The warm temperatures were accompanied by rain. Overnight as the storm developed, the rain turned to snow and most of the province experienced blizzard conditions. The snowfall set a new January record for Edmonton, with 35 centimetres of accumulation over the January 29 to 30 period. Less snow fell outside the Edmonton area, but temperatures across most of the province were well into the -30° C range, and wind chill temperatures throughout central and southern Alberta remained in the -35° to -50° C interval for several days.

*The darker the ice in an ice core, the colder temperatures were at the time the ice formed.*

## Spring Snowstorms

Just as we start to think of the coming warm weather and the pleasures of gardening, boating or returning to the golf course, Old Man Winter sometimes throws us a nasty surprise—the late-spring snowstorm. David Phillips, Canada's most well-known climatologist, has come up with the term "spring whitewash" as an appropriate description. These storms do not happen every year in Alberta, and they are not more common in any particular part of the province.

A storm that struck south-central Alberta in March 1988 was typical of the late spring snowstorm. On March 27, a low pressure system in the mid-troposphere moved from northern British Columbia into southern Alberta. As it crossed the Continental Divide, its motion slowed and the associated circulation at the surface intensified. Moist air

## Air Flow Pattern in the Middle Atmosphere on March 28, 1988

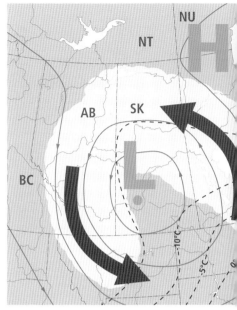

Fig. 8-1

# Snowfall Amounts for March 26–28, 1988

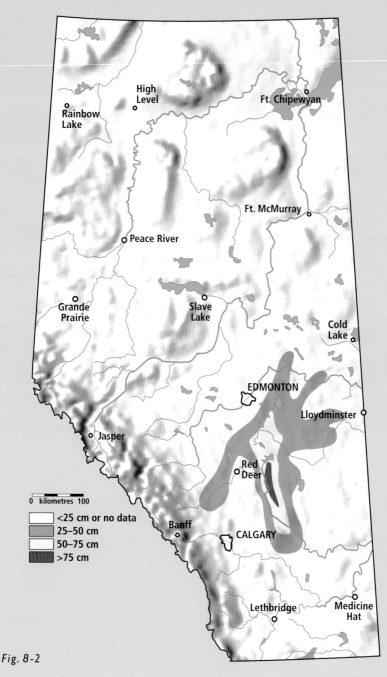

Legend:
- <25 cm or no data
- 25–50 cm
- 50–75 cm
- >75 cm

*Fig. 8-2*

rotating counter-clockwise around a low pressure system is usually ascending, and a substantial layer of saturated air was also moving toward the Rocky Mountains. Figure 8-1 illustrates the circulation pattern for the storm, showing the cloud band that was wrapped around the storm system. As this moist air moved upslope toward the Continental Divide, heavy precipitation began in the northwest quadrant of the low. With this storm, near-blizzard conditions lasted for 7 to 12 hours in south-central Alberta. The storm deposited a substantial band of snow through the middle of the province (see Figure 8-2). Heavy snowfall occurred through south-central Alberta, with Drumheller, Stettler and Camrose reporting more than 50 centimetres. The greatest snowfall amount reported was from just south of Stettler in the hamlet of Scollard, where 82 centimetres fell from March 26 to 28. Transportation was brought to a standstill with the heavy, wet snow. On the upside, however, those areas that were hit with the heavy snowfall benefited from substantial crop growth in spring, whereas areas 80 to 100 kilometres away suffered because crops were severely moisture stressed.

## Alberta Clippers

Rapidly moving snowstorms that form in Alberta and move off to the east are called Alberta Clippers. Eastern North America generally bears the brunt of this export of our natural resource. These storms are a lower tropospheric manifestation of a rapidly moving disturbance in the upper atmosphere westerly flow pattern. The disturbances move eastward from the Gulf of Alaska, crossing the continent in a matter of two or three days. When the upper level disturbance crosses the mountains, the associated surface low pressure storm system becomes somewhat disorganized. But in the lee of the Rocky Mountains, the upper level disturbance initiates the development of a low pressure cyclone in Alberta. This low usually does not have a lot of moisture associated with it as it moves out of Alberta to the southeast. As the Alberta Clipper speeds by, it drops 10 or so centimetres of snow on any location that it passes. Sometimes these storms intensify as they cross the Great Lakes or move along the east coast of North America, and they can then cause much more grief than they inflict on Alberta.

## The Storms of Summer

Thunderstorms are an almost daily occurrence across Alberta in summer. They occur most frequently through the centre of the province (see Figure 8-3). Thunderstorms are the source of virtually all severe summer weather, including hail, torrential rain and flash floods, damaging winds and tornadoes. Even weak thunderstorms are a threat because of lightning. In fact, lightning kills more people than any other summer weather hazard.

### Tornadoes

The fury of tornadic storms is unmatched by any other weather phenomenon. The violence of tornadoes is concentrated, and wind speeds can reach over 400 kilometres per hour, though no reliable measurements have been taken. Conventional instrumentation could not withstand the pummelling tornadoes would dish out, even if it happened to be situated at the precise location where these

relatively small-scale events occurred. Tornado strength estimates are based on the nature of the damage these storms inflict on structures. A tornado strength scale developed by Tetsuya Fujita in the late 1960s is still widely used to estimate tornado intensity.

Fortunately, tornadoes are rare occurrences in Alberta. The number of tornadoes varies widely from year to year, and the long-term Alberta average is about 10 per year. Tornadoes have been reported in all regions of the province, but because they do not last long, we can sometimes only conclude that one touched down by the trail of destruction it left behind. Tornadoes are rarely reported at weather observation stations, and most of the data on tornadoes comes from reports from the public and from volunteer severe weather watchers. As illustrated in Figure 8-4 tornadoes can be expected most frequently in an area extending from south of Edmonton to southeast of

## Average Annual Number of Days with Thunderstorms

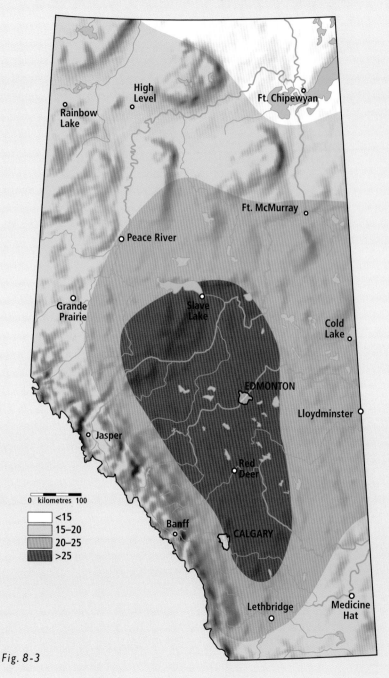

Rainbow Lake

High Level

Ft. Chipewyan

Ft. McMurray

Peace River

Grande Prairie

Slave Lake

Cold Lake

EDMONTON

Lloydminster

Jasper

Red Deer

0  kilometres 100

<15
15–20
20–25
>25

Banff

CALGARY

Lethbridge

Medicine Hat

*Fig. 8-3*

Lethbridge. The maximum of 0.82 tornadoes per 10,000 square kilometres per year near Edmonton is about one-fifth of the North American peak, which occurs in Oklahoma. A study by Dr. Keith Hage, professor emeritus at the University of Alberta, looked at the long-term incidence of tornadoes in Alberta. His study found that strong tornadoes were more frequent during the early part of the 20th century. From the late 1980s to 2000, the number of tornado sightings per year averaged in the mid- to high 20s, but in the last few years, the number of sightings appears to have dropped. In Alberta, there were only five reports of tornadoes in 2006 and 16 in 2007.

What meteorological conditions give rise to the formation of tornadoes? Tornadoes are associated with fierce thunderstorms. For intense thunderstorms to form, a number of atmospheric conditions need to come together in the correct sequence and at the correct time and location. Most important is the availability of a stream of warm, humid air in the lower atmosphere. Also, the middle to upper atmosphere should be dry and relatively cool. This temperature condition gives rise to what is known as convective instability. With a convectively unstable atmosphere, air parcels that rise from near ground level are warmer than their surrounding environment and continue to move upward. Strong solar heating is also required, but a layer of warmer air above the surface that results in a thermal inversion should be present to suppress the convection until later in the day. Essentially, the inversion acts like a lid on a boiling pot of water—when the steam builds up, the lid blows off. Much the same thing happens with the atmosphere. The inversion inhibits the early formation of cumulus clouds and small thunderstorms that would dissipate much of the energy. Movement of colder air aloft

## Three Types of Weather Alerts

**Special Weather Statements**
- issued for unusual weather events that could inconvenience or alarm the general public and that are not described in the weather forecast
- may be issued for events that are occurring outside of but may make their way into the forecast region
- updated as needed

**Watches**
- issued when conditions are right for a potentially disruptive storm, but when the track and strength of the storm are unknown.
- in summer, may be issued up to six hours before the event
- in winter, issued at least 12 to 24 hours before the event

**Warnings**
- issued when severe weather is occurring or will occur
- most are issued 6 to 24 hours in advance, except thunderstorm warnings, which may be issued less than one hour before the event
- are updated at least every six to eight hours or as needed

# Average Annual Frequency of Tornadoes per 10,000 Km²

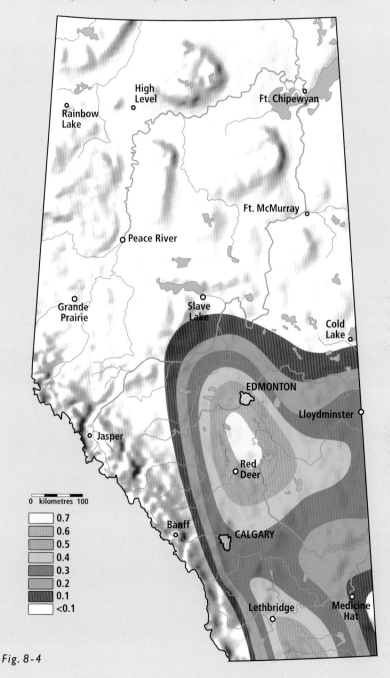

*Fig. 8-4*

The Edmonton tornado of July 31, 1987.

over the warm, moist lower level air is favourable to the formation of intense updrafts. Such a condition is usually associated with an advancing cold front. And ahead of the cold front, an upper level jet stream adds an additional stimulus for thunderstorm development. Portions of the jet stream have areas of ascent that can help to destabilize the upper atmosphere. As well, winds increasing with height tilt the storm cell and foster its continued growth.

The most destructive tornado to hit Alberta was the Black Friday storm that struck Edmonton on July 31, 1987. This was by far the most severe of the 12 tornadoes that have hit the Edmonton area since 1890. Twenty-seven people died, and approximately 250 people were injured. Property damage amounted to more than $250 million. The thunderstorm that spawned the tornado was growing explosively ahead of a cold front advancing across the province from the west. The thunderstorm cell that spawned the tornado was moving to the north, caught in the upper level southerly winds. The tornado first touched down near Leduc and tracked north for a full hour, traversing a distance of approximately 40 kilometres. As with most tornadoes, the intensity varied along its path. Based on damage reports, the tornado was classified as a strong F3 on the Fujita Scale, though it occasionally decreased to F0 and approached F4 at the mid-point of its path. The width of the damage zone also varied, ranging from 100 to as many as 1000 metres. The tornado demonstrated a characteristic common to most severe tornadoes—the presence of smaller funnels rotating around the main funnel. These so-called suction vortices are often responsible for much of the tornado damage.

The next most severe Alberta tornado occurred at the Pine Lake campground, south of Red Deer, on July 14, 2000. Twelve people lost their lives, and an additional 140 were injured. A thunderstorm supercell moving eastward with an advancing cold front spawned the tornado. This tornado moved along the ground at an average speed of 45 kilometres per hour, with wind speeds estimated to be up to 300 kilometres per hour, based on damage assessments. The tornado had a total track length of about 20 kilometres and reached a maximum intensity of F3.

Tornadoes in Alberta are usually formed in the high summer season, from late June to early August. Because thunderstorms often reach their peak intensity late in the day, the associated tornadoes usually happen between 4 and 7 PM. However, tornadoes can occur at any time of the day, as long as the conditions are right. The atmospheric control features for thunderstorms are usually changing. When they are all aligned, the intensity of thunderstorm cells increases, and when they are not aligned, the intensity decreases. A moving thunderstorm can spawn a series of tornadoes as the thunderstorm varies in intensity. This occurred on June 30, 1982, with a thunderstorm that initially developed in the foothills around noon. As the thunderstorm moved east-northeast near Rocky Mountain House, it produced its first tornado, which moved along a 22-kilometre track in mid-afternoon and attained a strength estimated at F2. As the storm moved eastward, it produced another tornado

around 6:00 PM near Lacombe, and one south of Viking around 9:20 PM.

Another tornadic storm that had a long track was particularly memorable for me. During the afternoon of June 29, 1984, a thunderstorm formed north of Rocky Mountain House and started moving to the northeast. The first sighting of a funnel cloud and possible tornado was made near Stony Plain. The thunderstorm continued moving to the northeast and spawned a weak tornado, which caused minor damage just northeast of the town of Athabasca around 4 PM. Thirty or so minutes later, a tornado with an estimated intensity of F1 passed south of the town of Calling Lake. About 20 kilometres north of the town, one individual was killed when his house collapsed on him. As the tornado moved farther north, two more farmsteads were damaged and a new home was destroyed before the twister retreated into its parent thunderstorm cloud. The thunderstorm system continued moving northeast, and a waterspout was seen over the water surface of Calling Lake. Near midnight, as the thunderstorm moved past Fort McMurray, one individual was killed by lightning. This thunderstorm system had travelled more than 400 kilometres over a 10-hour period, producing at least three tornadoes. I did a storm survey of the area south of Calling Lake after the storm, and the fickle nature of tornado damage was remarkable. The tornado moved down a shallow valley near the hamlet of Big Coulee, snapping off trees 0.5 metres in diameter. A new bungalow was completely destroyed in the hamlet itself, while only 50 metres away from the house, a small decades-old church did not even suffer shingle damage.

# Alberta Climate Extremes 1971–2000

## BANFF
| | |
|---|---|
| MAXIMUM (° C) | 34.4 ON JULY 28, 1934 |
| MINIMUM (° C) | -51.2 ON JANUARY 25, 1950 |
| DAILY RAINFALL (MM) | 53.1 ON JULY 13, 1948 |
| DAILY SNOWFALL (CM) | 58.6 ON DECEMBER 14, 1979 |
| SNOW DEPTH (CM) | 132.0 ON MARCH 1, 1972 |

## CALGARY
| | |
|---|---|
| MAXIMUM (° C) | 36.1 ON JULY 15, 1919 |
| MINIMUM (° C) | -45.0 ON FEBRUARY 4, 1893 |
| DAILY RAINFALL (MM) | 95.3 ON JULY 15, 1927 |
| DAILY SNOWFALL (CM) | 48.4 ON MAY 6, 1981 |
| SNOW DEPTH (CM) | 38.0 ON MARCH 18, 1998 |

## COLD LAKE
| | |
|---|---|
| MAXIMUM (° C) | 36.1 ON JULY 12, 1964 |
| MINIMUM (° C) | -48.3 ON JANUARY 20, 1954 |
| DAILY RAINFALL (MM) | 93.7 ON JUNE 5, 1962 |
| DAILY SNOWFALL (CM) | 41.8 ON APRIL 20, 1985 |
| SNOW DEPTH (CM) | 74.0 ON MARCH 15, 1974 |

## EDMONTON
| | |
|---|---|
| MAXIMUM (° C) | 34.5 ON AUGUST 5, 1998 |
| MINIMUM (° C) | -48.3 ON DECEMBER 28, 1938 |
| DAILY RAINFALL (MM) | 114.0 ON JULY 31, 1953 |
| DAILY SNOWFALL (CM) | 39.9 ON NOVEMBER 15, 1942 |
| SNOW DEPTH (CM) | 84.0 ON MARCH 18, 1974 |

## FT. CHIPEWYAN
| | |
|---|---|
| MAXIMUM (° C) | 34.4 ON JUNE 4, 1970 |
| MINIMUM (° C) | -50.0 ON JANUARY 26, 1969 |
| DAILY RAINFALL (MM) | 71.8 ON JUNE 19, 1986 |
| DAILY SNOWFALL (CM) | 27.4 ON OCTOBER 12, 1984 |
| SNOW DEPTH (CM) | 124.0 ON MARCH 18, 1974 |

## FT. McMURRAY
| | |
|---|---|
| MAXIMUM (° C) | 37.0 ON AUGUST 10, 1991 |
| MINIMUM (° C) | -50.6 ON FEBRUARY 1, 1947 |
| DAILY RAINFALL (MM) | 94.5 ON AUGUST 6, 1976 |
| DAILY SNOWFALL (CM) | 29.7 ON MARCH 16, 1951 |
| SNOW DEPTH (CM) | 68.0 ON JANUARY 13, 1996 |

## GRANDE PRAIRIE
| | |
|---|---|
| MAXIMUM (° C) | 34.5 ON AUGUST 9, 1981 |
| MINIMUM (° C) | -52.2 ON JANUARY 2, 1950 |
| DAILY RAINFALL (MM) | 90.0 ON AUGUST 5, 1994 |
| DAILY SNOWFALL (CM) | 36.3 ON OCTOBER 4, 1957 |
| SNOW DEPTH (CM) | 86.0 ON FEBRUARY 1, 1982 |

## HIGH LEVEL
| | |
|---|---|
| MAXIMUM (° C) | 35.2 ON AUGUST 9, 1981 |
| MINIMUM (° C) | -50.6 ON JANUARY 13, 1972 |
| DAILY RAINFALL (MM) | 103.4 ON JULY 13, 1998 |
| DAILY SNOWFALL (CM) | 42.6 ON MAY 20, 1989 |
| SNOW DEPTH (CM) | 104.0 ON MARCH 19, 1987 |

## JASPER
| | |
|---|---|
| MAXIMUM (° C) | 36.7 ON JULY 16, 1941 |
| MINIMUM (° C) | -46.7 ON JANUARY 19, 1935 |
| DAILY RAINFALL (MM) | 107.7 ON AUGUST 5, 1969 |
| DAILY SNOWFALL (CM) | 51.6 ON FEBRUARY 17, 1948 |
| SNOW DEPTH (CM) | 94.0 ON FEBRUARY 1, 1974 |

## LETHBRIDGE
| | |
|---|---|
| MAXIMUM (° C) | 39.4 ON JULY 10, 1973 |
| MINIMUM (° C) | -42.8 ON JANUARY 3, 1950 |
| DAILY RAINFALL (MM) | 85.4 ON MAY 23, 1980 |
| DAILY SNOWFALL (CM) | 55.1 ON SEPTEMBER 21, 1968 |
| SNOW DEPTH (CM) | 86.0 ON APRIL 30, 1967 |

## LLOYDMINSTER
| | |
|---|---|
| MAXIMUM (° C) | 37.4 ON AUGUST 5, 1998 |
| MINIMUM (° C) | -42.5 ON FEBRUARY 7, 1994 |
| DAILY RAINFALL (MM) | 63.4 ON JULY 29, 1986 |
| DAILY SNOWFALL (CM) | 57.6 ON APRIL 27, 1991 |
| SNOW DEPTH (CM) | 49.0 ON JANUARY 29, 1997 |

## MEDICINE HAT
| | |
|---|---|
| MAXIMUM (° C) | 42.2 ON JULY 12, 1886 |
| MINIMUM (° C) | -46.1 ON JANUARY 22, 1886 |
| DAILY RAINFALL (MM) | 121.9 ON AUGUST 14, 1927 |
| DAILY SNOWFALL (CM) | 33.8 ON MARCH 11, 1954 |
| SNOW DEPTH (CM) | 50.0 ON NOVEMBER 26, 1990 |

## PEACE RIVER
| | |
|---|---|
| MAXIMUM (° C) | 36.7 ON JULY 13, 1945 |
| MINIMUM (° C) | -49.4 ON JANUARY 13, 1950 |
| DAILY RAINFALL (MM) | 53.0 ON MAY 4, 2000 |
| DAILY SNOWFALL (CM) | 26.2 ON APRIL 12, 1966 |
| SNOW DEPTH (CM) | 117.0 ON APRIL 1, 1967 |

## RED DEER
| | |
|---|---|
| MAXIMUM (° C) | 36.0 ON AUGUST 5, 1998 |
| MINIMUM (° C) | -43.3 ON DECEMBER 9, 1977 |
| DAILY RAINFALL (MM) | 97.4 ON JULY 29, 1981 |
| DAILY SNOWFALL (CM) | 28.5 ON SEPTEMBER 17, 1997 |
| SNOW DEPTH (CM) | 75.0 ON MARCH 3, 1997 |

## SLAVE LAKE
| | |
|---|---|
| MAXIMUM (° C) | 33.3 ON JUNE 3, 1970 |
| MINIMUM (° C) | -42.8 ON JANUARY 30, 1971 |
| DAILY RAINFALL (MM) | 82.4 ON JUNE 25, 1983 |
| DAILY SNOWFALL (CM) | 25.9 ON MARCH 14, 1974 |
| SNOW DEPTH (CM) | 104.0 ON MARCH 18, 1974 |

Alberta's largest documented hailstone fell 8 km southeast of Wetaskiwin on July 6, 1975. It weighed 249 grams.

## Hailstorms

Hail is a regular companion to thunderstorms in Alberta, and a research program to investigate hailstorms operated in the province for three decades. As shown in Figure 8-5, hailstorms are most frequent in west-central Alberta. The number of hail days peaks in the summer months, and on average, hail occurs on more than 21 days in July. The most severe thunderstorm events are associated with significant quantities of hail, and some of the stones in these storms can reach large sizes. During the Edmonton tornado of July 1987, hail was reported across most of the city.

*Hail comes only from convective clouds, usually of the cumulonimbus variety.*

Hailstones ranged from the size of a walnut (about 3.5 centimetres across) to as large as a tennis ball (6.5 centimetres). With the Pine Lake tornado, reports of golf-ball–sized hail (about 4.5 centimetres across) were widespread, and areas well downstream of the tornadic event reported baseball-sized stones (7 to 8 centimetres across).

A major hailstorm occurred in Calgary on September 7, 1991. Damage claims from hail exceeded $300 million, ranking this storm as the second most costly weather event in Canadian history.

Hail events combined with heavy rain can cause flooding. In Edmonton on July 11, 2004, hail accompanied heavy rain from thunderstorms. The hail plugged storm sewer drains, and motorists were stranded as low areas on streets in Edmonton were flooded.

## Average Annual Number of Days with Hail

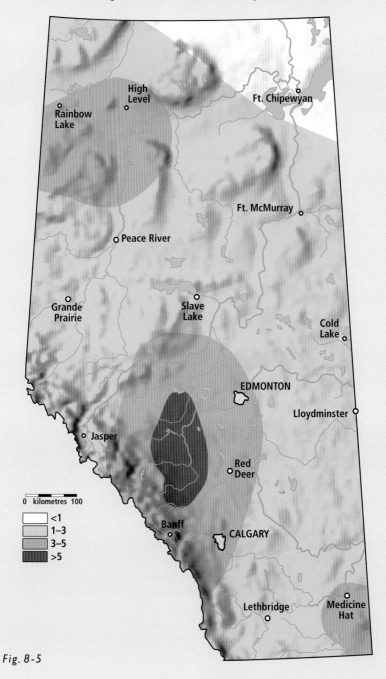

High
Level

Rainbow
Lake

Ft. Chipewyan

Ft. McMurray

Peace River

Grande
Prairie

Slave
Lake

Cold
Lake

EDMONTON

Lloydminster

Jasper

Red
Deer

0    kilometres   100

Banff

CALGARY

<1
1–3
3–5
>5

Lethbridge

Medicine
Hat

*Fig. 8-5*

## Heavy Summer Rainstorms

Extreme rainfalls occasionally besiege Alberta. The province's period of peak rainfall is early summer, when snowmelt in the mountains is at its maximum. With these heavy rainstorm events, the precipitation is usually prolonged over two or three days. These storms are usually associated with what is known as a cold low. Cold lows are nearly circular and can be several hundred kilometres across. The temperature pattern associated with these systems gives them their name—the coldest temperatures are near their centres. Cold low systems differ from the rapidly moving low pressure systems that travel along in the westerlies. Those systems usually have warm air located near their centres. The atmospheric circulation is similar to that which occurs with spring snowstorms, as described earlier in this chapter. Again, the Rocky Mountains play a role with these storms, especially if the moist air circulating around the northern portion of the low is forced to ascend as it moves toward the mountains. In late spring and early summer, this can result in intense flooding if heavy rains fall on the headwaters of a drainage basin that is already swollen with snowmelt and runoff.

*Most of the precipitation that falls in Canada draws its water from the Pacific Ocean, the Gulf of Mexico and the Caribbean Sea.*

One particularly heavy summer rainstorm occurred in southern Alberta on June 6 and 7, 1995. Over a three-day period, the cold low in the upper atmosphere crept across the northwestern states of Washington, Idaho and Montana. Although the surface low pressure system was located in Montana, well to the south of Alberta, a strong easterly circulation prevailed over the southern half of the province for two to three days. As is often the case with these cold low situations, daytime heating triggered the formation of towering cumulus clouds and thunderstorms ahead of the advancing cold low. These clouds produced scattered showers on June 5 and by midnight had evolved into widespread rain. Heavy rain continued the next day, finally ending by noon on June 7. The area stretching from Lethbridge to just west of Calgary received more than 50 millimetres of precipitation, and many locations in the foothills received more than 160 millimetres.

## Dust Storms

Albertans are often faced with dry conditions that are the result of a region getting below-normal precipitation for a lengthy period of time. In dry conditions, strong winds often stir up dust storms, obscuring visibility and causing soil erosion. Dust storms are frequent in the desert areas of the Middle East, where they are called haboobs (derived from the Arabic term for violent wind). They are fairly common in the semi-desert areas of Arizona and were often reported in western Canada during the dust bowl years of the 1930s. At the Namao Air Show on May 23, 1988, tens of thousands of people were surprised by a sudden, widespread dust storm that swept through central Alberta. It had all the appearance of the haboob. Dry conditions had prevailed through much of south and central Alberta in the 1980s. During the winter of 1987–88, the Edmonton area had received about 40 percent less precipitation than usual, and

temperatures during winter and spring were averaging 4° C above normal. So, by the Victoria Day long weekend, conditions were extremely dry.

*The foothills region west of Edmonton that stretches to the Rocky Mountains is a lightning hot spot.*

During the Namao storm, a line of thunderstorms developed along an advancing cold front. The circulation structure in the upper atmosphere was also in a configuration to provide the necessary conditions for a severe weather event. In fact, the atmosphere had many of the characteristics seen during the Black Friday tornado episode less than a year earlier. All that was missing was moist air at low levels. The thunderstorms along the advancing cold front had relatively high bases because the low level air feeding them was dry. Such dry conditions, however, also contribute to the formation of strong downdrafts from thunderstorms. As precipitation falls from the base of these thunderstorm cells, there is a great deal of evaporation, which takes heat from the air entrained by the falling precipitation. Although descending air normally warms, the evaporative cooling makes descending air parcels become negatively buoyant relative to the surrounding air. The parcels continue to accelerate downward, and a strong downdraft, often known as a microburst, is associated with the thunderstorm. These strong downdraft winds stir up the dry surface soils and create a dust storm. Figure 8-6 depicts the wind flow that was associated with the dust storm. During the 1988 event, a dust band was created that was about 100 kilometres wide and about 300 kilometres long. The dust cloud was around 600 metres deep, and visibility was reduced to less than 400 metres over a wide area, dropping to near zero on occasions.

## Dust Storm Wind Pattern

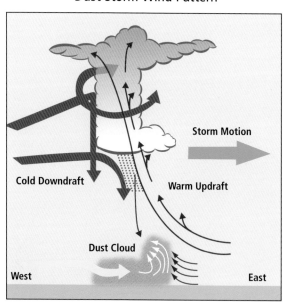

**Storm Motion**

**Cold Downdraft**

**Warm Updraft**

**Dust Cloud**

**West**

**East**

*Fig. 8-6* Thunderstorms supporting the storm had bases 4 km above ground. Downdrafts accelerate toward the surface, creating a microburst and associated dust cloud.

# Chapter Nine:
# Observing the Weather

**Old weather saying:**
*Rain before seven,
fine by eleven.*

What will the weather be like tomorrow? Next Friday? This time next year? These are questions that we ask. Looking out the window gives us our first impression as to what the weather will do for the next few hours, but if we are interested in what is going to happen tonight or tomorrow, that simple process is inadequate. In the 19th century, when people first undertook an intensive, organized study of the atmosphere, they realized there could be very different conditions at two locations, even when the locations were not far apart. This realization led to the understanding that weather is organized into patterns and that these weather patterns move. So it was concluded that observations should be taken at many locations at about the same time and that the observations should be gathered together at a central location where the patterns could be analyzed. Then a forecast of future weather conditions could be prepared based on an understanding of the rules or laws that determine how the atmosphere works.

The nature of weather and how it behaves on different scales is based on the principles of fluid dynamics. The atmosphere is a fluid that is in turbulent flow on a rotating sphere. Mathematician Lewis Fry Richardson may have summarized it best when, early in the 20th century, he described turbulent fluid flow as "big whirls have little whirls that feed on their velocity, and little whirls have lesser whirls and so on to viscosity." This ditty recognizes some basic realities of the atmosphere: that it is a turbulent fluid that behaves on many different scales of motion, and that there is a link between these scales of motion.

*Because measurements of weather parameters are taken from all over the world, the information is recorded in a special code that can be understood by the people of all countries, regardless of the language they speak.*

So, the essence of weather forecasting is to measure the smallest scales reasonable and use the rules of atmospheric motion that define the link between these scales to give a picture of the state of the fluid at some time in the future. Weather patterns have a space scale of a few hundred kilometres, and they can change over a few hours. To resolve these weather patterns requires a network of stations spaced about 100 kilometres apart that take observations every hour. This is the space and time scale of most meteorological observing programs that support weather forecasting.

# Observation Systems

Assembling observations and preparing forecasts for even a few locations requires a large-scale, coordinated approach. Consequently, the observing and forecasting systems are largely the responsibility of national organizations. The data-observing approach must be standardized to make the international exchange of data run smoothly. Observational standards follow an international protocol established and maintained by the World Meteorological Organization, which operates under the United Nations. The data is communicated to one or more locations, where it can be analyzed so forecasts can be prepared.

Weather forecasting requires data from the full depth of the atmosphere. In Canada, the surface weather observation network consists of several hundred stations across the country that repeat measurements, generally on an hourly basis. In addition, about 30 stations measure meteorological variables above the surface to heights of about 30 kilometres. At these aerological stations, surface weather data is also measured, and many other specialized instruments are used for specialized observations that support atmospheric research programs. These surface and upper air in-situ measurements are supplemented with data from remote sensing systems such as weather radar and weather satellites. The data is sent to local and national weather offices, and much of it is made available to the public through the Internet and the media.

## Surface Weather Observation

Formal weather observations were recorded at locations across British North

An automated weather observing station at Elk Island National Park

America through the 18th and early 19th centuries. The first official observation program began on Christmas Day in 1839. According to a plaque on the University of Toronto campus, the British Army began regular meteorological and magnetic observations on the site, then called Her Majesty's Magnetical and Meteorological Observatory, in 1840. John Lefroy, the observatory's director, travelled west in 1843 to make magnetic observations in the Hudson's Bay Company's territories. His weather observations at Fort Chipewyan and Fort Edmonton were among the earliest formal measurements taken in Alberta. The first weather observation network was established in Great Britain in the 1850s. Canada's first network was established in 1872. It covered southern Ontario, Quebec and the Maritime provinces, and the data was communicated using the newly installed telegraph systems. During the 1880s, the network expanded into what is now Alberta, following the Canadian

Pacific Railway and its telegraph network. At present, the Meteorological Service of Canada operates the national weather observation program that supports the preparation of weather forecasts.

The measurement of atmospheric variables has a long history, and electronic technology and computerization have influenced weather observing methodologies. Meteorology has evolved into several areas of specialization, and the monitoring requirements have evolved as well. Data requirements for weather forecasting are somewhat different than those needed for atmospheric research or for specialized applications in air quality. Weather forecast requirements are based on what are known as standardized observing protocols. Issues including instrument location, representativeness and accuracy of measurement, and observation averaging times must be considered if data is to be used for the forecast. Predicting the atmosphere is difficult enough without the need to take into account idiosyncrasies for each weather measurement.

At weather observing stations, the instruments are located where they provide a measurement that is representative of the atmosphere. This means, for example, that they are not sheltered by a tree or building, that a temperature reading is not artificially elevated by direct sunlight or the heat rising off a paved surface, that snow measurement is not subject to a snowdrift depending on wind direction, and that a precipitation measurement is not elevated by rain dripping off a roof or swirling around a tree.

Most observations are from unstaffed sites where data is measured automatically. Many sites are located at airports, where trained observers provide additional data that is useful for the aviation industry.

Many instruments are housed in a ventilated cabinet known as a Stevenson Screen. Named after its inventor Thomas Stevenson (father of the famous author Robert Louis Stevenson), the cabinet ensures that sunlight does not fall directly on temperature gauges. The cabinet is painted white to ensure that the sun's rays do not cause the cabinet to heat.

Each Stevenson Screen contains different types of thermometers.

## Precipitation

The history of precipitation measurement goes back to the 4th century BC in India. The principle used then is the same one used now—measure the amount of precipitation collected in an open-topped container. Refinements have been focused on reducing the loss of precipitation from splash-out of rain drops, reducing the effects of wind on collection efficiency of the opening, and reducing evaporative loss between precipitation events. The manual rain gauge used in Canada has a design that has been largely unchanged since 1871. The rate of rainfall is measured by a tipping bucket rain gauge. Collected rain passes into a small container that empties each time its capacity is reached. The rate at which the container empties indicates the rainfall rate.

Snow measurement has proven to be a challenge. The snow catch rate of a collection container is greatly influenced by wind, so most devices have a shield to reduce wind speed around the opening. The shape of the opening also influences how efficiently a container collects snow. The Canadian Nipher-shielded Gauge has a bell-shaped opening to minimize loss. Where there is a human observer, the depth of snow accumulated on the ground is determined by a simple depth measurement at a number of sites in a so-called snow course. Measurements are taken at several locations so differences in accumulation that result from wind drifting can be averaged.

A tipping bucket rain gauge (above); the tipping bucket (below)

Many innovative techniques have been used to measure snowfall automatically. Snow pillows laying on the ground measure the weight of the snowpack that accumulates during winter. The storage gauge collects snow in a large container filled with glycol, which melts the collected snow, and measurements of the weight of the container are taken periodically. The depth of snow can also be measured at one point on the ground by a downward-pointing microwave sensor.

Automated detection that identifies the type of precipitation is difficult. One approach measures the fall rate of the precipitation by means of a small Doppler radar and links the fall rate to the type of hydrometeor. This works well enough to differentiate between snow and rain, but discriminating between drizzle and snow is not easy because they have essentially the same terminal velocity. The detection of freezing precipitation is also a challenge. One approach uses a vibrating probe and calibrates the rate of vibration to the rate at which precipitation adheres.

### Temperature

Around 1592, Galileo invented the first thermometer, which used the expansion of air to measure temperature. However, that approach was prone to many errors, and the apparatus was large and cumbersome. Scientists soon realized that thermometers made of sealed tubes containing mercury or alcohol were much more reliable and practical.

Gabriel Daniel Fahrenheit developed a temperature scale in 1724 that had three fixed reference points. Zero was the value given to the coldest temperature that he could create by mixing ice, water and ammonium chloride. The upper end of the scale was the temperature of the human body, which Fahrenheit arbitrarily called 96 degrees. The third fixed point was the temperature of a water/ice mixture that, on his scale, had a value of 32 degrees. The Celsius temperature scale, invented by the Swedish astronomer Anders Celsius in 1742, is the one most frequently used worldwide. On this scale, the temperature of the water/ice

Microwave snow-depth sensor (above); Nipher-shielded Gauge (left)

Maximum and minimum thermometers in a Stevenson screen

mixture was set as zero, and the value of 100 degrees was assigned to the temperature at which pure water boils.

Various types of thermometers are used at surface weather stations. The minimum thermometer contains a tiny metal dumbbell (index) that floats in liquid, usually alcohol. The index is forced toward the bulb by the retreating surface of the liquid as the temperature falls. When the temperature rises, the dumbbell remains in place and registers the minimum temperature. The maximum thermometer has an index that is pushed upward in the barrel of the thermometer and registers the maximum value reached in the observation period. Regular, maximum and minimum thermometers are all manually read.

Temperature sensors have been developed for use in automated systems or where space limitations or working environments are unsuitable for liquid-filled thermometers. The most common are instruments known as thermistors (the name is a combination of "thermal" and "resistor"), which are made of mixed metal oxide semiconductors. In a semiconductor, the resistance to the flow of electrical current changes as the temperature of the device changes. In some thermistors, the resistance increases with increasing temperature, and in some, the resistance decreases with increasing temperature. The semiconductor devices can be physically small and rugged in design, which makes them suitable for use with electronic circuitry.

161

A sling psychrometer to measure humidity

## Humidity

Humidity measurements are important in meteorology because knowledge of the moisture content of air is necessary to determine the formation of cloud. In 1660, Francesco Eschinardi demonstrated that the evaporation of water causes a thermometer to cool. The psychrometer is a device that uses this principal to determine the amount of moisture in the air based on the difference in temperature between two thermometers. The dry bulb thermometer measures air temperature normally, whereas the so-called wet bulb thermometer has its bulb enclosed by a water-saturated wick. A fan draws air past both thermometers at a known rate, and the temperature difference is then used to determine the humidity based on previously calibrated values.

The psychrometer is not suitable for automated measurements of atmospheric humidity because the water reservoir would need to be maintained, and the behaviour of the device changes at temperatures below freezing. An instrument called the dewcel has been developed for use at Canadian weather stations. This device has a fibreglass sleeve that is saturated with a lithium chloride solution. The lithium chloride absorbs moisture from the air, and a heater is used to raise the temperature of the sleeve, causing the moisture to evaporate. A thermistor measures the temperature difference between the heated and unheated sleeve. This temperature difference is proportional to the atmospheric moisture.

Many other devices have been developed to measure humidity. They use different principles to measure moisture, such as the change in length of hair as it absorbs water vapour, the change in electrical properties of materials as they absorb water vapour, or the temperature at which a cooled reflective surface becomes fogged because of water vapour.

*Canada's Alert weather station is the world's most northerly. It is located on Ellesmere Island, 830 kilometres from the North Pole.*

## Pressure

Atmospheric pressure is a result of the weight of overlying air. Surface pressure has traditionally been measured by the mercury barometer (meaning "weight meter"), which Evangelista Torricelli invented in 1644. The modern instrument consists of a mercury-filled tube that is inverted and immersed in a mercury-filled cistern. Atmospheric pressure pushes down on the top of the cistern and supports the column of mercury in the tube. The height of mercury that is supported essentially balances the weight of the atmosphere above the cistern. At sea level, this amounts to the weight of about 76 centimetres of mercury. This is the most consistent and reliable type of atmospheric measurement available. However, the mercury barometer is not easily transportable, it does not lend itself to automation and there are health concerns associated with using mercury. As a result, the aneroid barometer, invented by Lucien Vidie in 1843, is widely used.

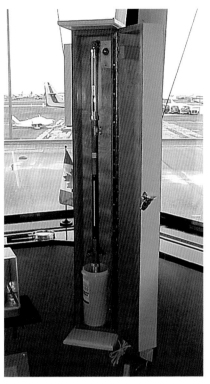

Mercury barometer at an Alberta airport (above); aneroid barometer with graph to record pressure changes (below)

The aneroid barometer consists of a partially evacuated and sealed metal bellows canister that expands or contracts in response to atmospheric pressure changes. Mechanical gears or electronic techniques determine the amount of this deflection.

*Flying insects stay closer to the ground when the air pressure drops, so when you see swallows flying low to the ground, it could be an indication that rain is on the way.*

Aneroid systems are used in portable barometers, altimeters and pressure recording devices (barographs). Other pressure measurement techniques using quartz and silicon-based cells are used in specialized applications. To measure atmospheric pressure, the barometer must be located in a building where the temperature is controlled for the stability of the instrument. To ensure the pressure represents outside conditions, the building must not be sealed, and it must be designed to minimize the effects of spurious pressure fluctuations caused by wind pressure, vehicles moving nearby and even the opening and closing of doors.

### Wind

Wind may be the most influential weather phenomenon throughout the ages. The Tower of the Winds in Athens was built by Andronicus of Cyrrhus around 50 BC to recognize the eight quadrants from which the wind blows. A weather vane was mounted on top of the tower. Since that time, various types of wind vanes have been developed to measure wind direction, and several different methods have been used to measure the magnitude of the wind. One early approach attempted to measure wind force directly using a suspended flat plate that was deflected by the wind. Such devices date back to about 1450. Subsequent variations on this theme consisted of a flat plate, either square or circular, which was kept facing the direction of the wind by a vane. One such device invented by A.F. Osler was installed at the observatory in Greenwich, England, in the mid-19th century.

Osler's wind measurement device

A more useful measurement concept is wind speed—the distance that air moves over a short period of time. Wind speed is often measured with the cup anemometer, which operates on the principle that cups mounted on arms rotate around a vertical axis as the air moves past. The number of rotations in a fixed interval of time is proportional to the wind speed. John Thomas Romney Robinson invented the first such instrument in 1846, and Canadian John Patterson invented the common three-cup anemometer in 1926. The anemometer and wind vane are generally mounted on a tower that meets an international standard for height and

The Robinson cup anemometer

siting. Measurements are taken 10 metres above ground, and the tower is located where the wind flow will not be influenced by any trees or buildings.

A standard wind vane with the Patterson three-cup anemometer

A propeller-and-vane anemometer

in 1892 is still used today. This approach measures the speed at which an aircraft moves through the air, but it is also used in ground-based instruments to measure rapid fluctuations in wind speed. The sonic anemometer calculates wind speed by measuring the time taken by two sound pulses to move in opposite directions across a known distance. The difference in time that the two pulses travel is proportional to the speed of the air. This instrument has the advantage of no moving parts and can measure rapid changes in wind speed.

> It was one of those March days when the sun shines hot and the wind blows cold: when it is summer in the light, and winter in the shade.
> —Charles Dickens,
> *Great Expectations*

Other wind measurement approaches have been developed for special applications. At many automated weather stations, the wind speed and direction devices are combined into one instrument. This device has a propeller mounted at the front of a wind vane. Non-mechanical devices are used to measure rapid fluctuations in wind speed and direction. Another approach uses instruments known as hot-wire and hot-film anemometers to measure the cooling effect of air as it moves across an electrically heated surface. A stronger wind removes more heat and, therefore, more electrical energy is required to heat the surface. Yet another approach measures the pressure drop that occurs when wind flows across a small opening. The pressure tube approach was first developed in the late 18th century, and a practical version developed by William Dines

### Visibility

This parameter is important for such things as aviation and marine transportation. Visibility is a function of atmospheric conditions, and it also depends on the capability of the human eye, making it one of the more subjective atmospheric measures. A number of instrumentation methodologies have been developed to bring a degree of objectivity to this parameter. Most instruments measure the drop in intensity of a light source when observed over a fixed distance in the atmosphere. These instruments are used at airports to measure the runway visual range, which is critical for aircraft movement around airports.

## Observations at Sea

A major international cooperative program is in place to gather weather information from over the oceans. Observations taken on vessels use much of the same types of instrumentation that are used at land stations. However, because the ships are moving, the wind measurements must take into account the vessel's motion. In addition, the ideal siting criteria used at land stations cannot be met in the cramped conditions onboard most vessels. The key measurements of temperature, pressure and wind are still useful in the weather forecasting programs.

A visibility sensor

Marine buoys are the oceanic version of the unstaffed automatic land stations. There are two types of buoys—moored or drifting. The moored buoys have a long tether fixed to the ocean floor. The observation program usually includes pressure, temperature and wind as well as marine measurements (such as sea surface temperature) and sea state information (such as wave period and height). Occasionally, the buoys also collect ocean current information. Drifting buoys are generally limited to temperature and pressure measurements, but some gauge wind speed, as well. Data is usually communicated through the meteorological satellites to the world communication network. Although the location of the moored buoys is known, the drifting buoys must be tracked using global positioning technology. The motion of the buoys gives a measure of the ocean current.

## Measurement Aloft—the Aerological Monitoring Program

Weather forecasting programs need regular measurements of the upper atmosphere. These aerological measurements, as they are known, are taken year-round every 12 hours, at noon and midnight Coordinated Universal Time (that's 6:00 and 18:00 Mountain Daylight Time in Alberta). The spacing of the upper-air observing network is much wider than the surface observing network, and stations are about 600 kilometres apart. In Alberta, one station in the national network is located at Stony Plain. Measurements are taken using a lightweight instrument package that is lifted using a helium-filled balloon.

Release of a radiosonde at the Stony Plain aerological station

The instrument package, known as a radiosonde, includes sensors that measure pressure, temperature and moisture. It has a small radio transmitter on board that sends the data and also allows the motion of the instrument to be tracked at the receiving ground station. The horizontal distance moved by the instrument package during its ascent is used to calculate the wind speed and direction. The balloon bursts at a height of about 30 kilometres above ground level, and a small parachute slows the instrument package in its fall back to earth. These instrument packages are not normally recovered. The balloon-borne observations are taken in all weather conditions. It is a particular challenge to release the balloon when the surface winds are strong and freezing rain or blowing snow is occurring.

Once a week, ozone measurements are taken at 10 Canadian aerological sites. This monitoring program focuses on ozone in the stratosphere and the depletion of this layer, which protects life forms from the harmful effects of ultraviolet radiation. At present, the Stony Plain site releases this ozonesonde. Ground-based instruments also measure ozone with remote sensing techniques.

Although the aerological network is confined primarily to land stations, there are a few ships that release radiosondes at fixed locations on the oceans. These programs are expensive to operate, and they are now being replaced by programs operated from ocean-going transport vessels, which release radiosondes while en route.

## Remote Sensing

Weather observations using instruments that are in direct contact with the atmosphere are known as in-situ measurements because they have sensors in direct contact with the atmosphere. Remote sensing techniques probe the atmosphere from a distance.

Various techniques are used for remote sensing, including sound, the reflection of electromagnetic waves (from sunlight, microwave and laser instruments) and the sensing of heat radiated by a substance.

## Weather Radar

When radar (an acronym for radio detection and ranging) was first used to track the motion of aircraft, the returns from clouds and precipitation were considered detractive noise. However, it quickly became apparent that radar could be used to provide new information about the atmosphere. Weather radars have a transmitting dish that rotates and sends pulses of radio waves in the microwave frequencies. The dish focuses the pulses along a narrow beam that reflects off cloud and water droplets.

When part of the microwave pulse bounces back to the antenna, the radar's electronics time the return and measure its intensity. One-half of the total time for the pulse to travel to and return from the target multiplied by the speed of light determines the target's distance from the radar dish.

Vulcan, AB radar dome

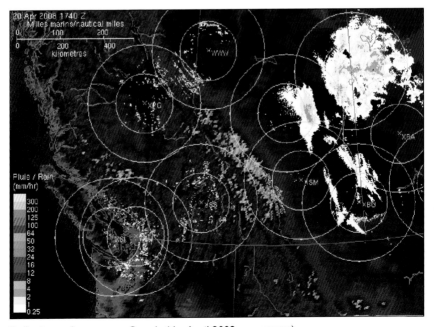

Radar image for western Canada (the April 2008 snowstorm)

**Weather Radar Interpretation Errors** (adapted from Environment Canada website)

**1. Blocking Beam**

Uneven terrain such as hills and mountains can block a radar beam, leaving gaps in the pattern—this is very common in the Rockies.

**2. Beam Attenuation**

Storms closest to the radar site reflect or absorb most of the radar energy, so little energy is left energy to detect storms that are farther away.

**3. Overshooting Beam**

When clouds are close to the ground, as they are in lake effect snowsqualls, the radar beam may overshoot them so that even clouds with intense precipitation will only have a weak echo.

**4. Virga**

The radar beam detects precipitation that is occurring aloft, but the precipitation never reaches the ground because it is absorbed by low-level dry air conditions.

**5. Anomalous Propagation**

During an inversion in the low atmosphere, when a layer of warm air overlies a layer of cooler air, the radar beam can't pass between the two layers and is reflected back to the ground, sending a false strong signal back to the radar site—this is most common in the morning.

**6. Ground Clutter**

The radar beam echoes off objects on the ground, such as tall buildings, trees or hills.

The strength of the reflected signal is proportional to the size of the reflecting object. Microwave radiation reflects off cloud and rain drops (and even swarms of insects). Based on the physical properties of raindrops, the intensity of the returned signal is proportional to the size distribution of the reflecting droplets. Using rain gauge data, the rainfall rates are correlated with the assumed droplet size. In this way, the rainfall rates in the sensed clouds can be estimated. Canadian weather radars transmit in the 5-centimetre wavelength and have an effective viewing distance out to about 250 kilometres. The radar display is a circular pattern centred at the radar site. Usually a sequence of images is displayed, which then shows the motion of the echo. The Canadian weather radar network is operated by Environment Canada and includes five sites in Alberta.

Some radars also make use of the Doppler principal to measure the shift in frequency of the returned echo. Changes in the frequency of the returned signal are proportional to the speed that the reflecting object is moving toward or away from the radar site. This Doppler principal is used to measure the flow patterns in the atmosphere. It is useful for examining the wind patterns in intense thunderstorms, and the Doppler radar can even be used to determine if the cloud echo is rotating. A rotating cloud echo usually indicates a severe thunderstorm, which may generate a tornado even though tornadoes themselves are below the resolution of the radar beam. The radar near Edmonton has Doppler processing capability, and there are plans to convert all radars so that they have this feature.

# The Doscpler Effect

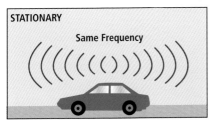

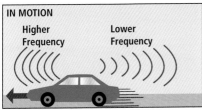

*Fig. 9-1* The frequency of sound waves is shifted because of the motion of the source. The frequency of microwave radar echoes is shifted because of motion of cloud and rain.

## Weather Satellites

The weather satellite is a relatively recent observational tool but has become one of the most useful tools for observing large portions of the globe. The TIROS satellite launched by the United States in 1960 was the first, followed by scores of satellites that have since been launched by governments and other agencies.

Most weather satellites use passive sensing systems, which receive two basic types of radiant energy. Visible light produced by the sun is reflected off the earth's surfaces and clouds, back up to the satellite. Visible sensors on meteorological satellites are essentially black and white cameras. Clouds look white to the satellite-based sensor, whereas ground surfaces and water bodies appear grey or black. A second type of sensor detects infrared or heat energy. The intensity of the infrared energy is related to the temperature of the emitting surface, a

phenomenon known as Boltzmann's law. The satellite infrared sensors are sensitive to the temperatures of earth's land and water surfaces as well as cloud tops—a temperature range extending from -70° to 60° C. Because the earth and atmosphere emit heat day and night, infrared images are always available, unlike the visible images, which are only available during daylight.

There are two types of satellite systems now generally in use. Geostationary satellites are placed in circular orbits over the equator at heights of 35,800 kilometres. The satellite moves in the same direction as the earth rotates and takes one full day to orbit the earth. It basically appears stationary above the earth's surface. The second type of satellite is in a much lower orbit at about 800 kilometres above the surface. The orbit passes near the poles and is at a slight angle of inclination relative to the equator. The orbital geometry is such that the

171

## Weather Satellites

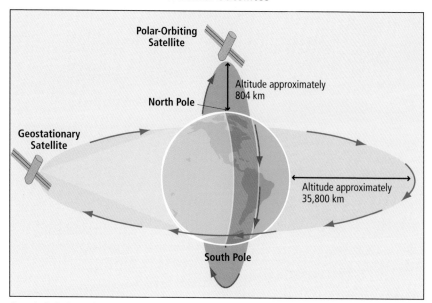

*Fig. 9-2* Polar-orbiting satellites are close to earth and provide high resolution images. Geostationary satellites are in high orbits and provide low resolution images.

satellite orbital plane moves in time with the sun. These polar-orbiting satellites take a continuous swath of images along their orbit and, as a consequence, the images appear as one long, continuous image. Because geostationary satellites are about 45 times higher than polar-orbiting satellites, the resolution is lower, and images from the equator-oriented satellites become greatly distorted toward the north and south poles. The polar-orbiting satellites move around the earth in about 100 minutes, and most places on the earth's surface are scanned twice daily, once in daylight and once in darkness.

A polar-orbiting satellite image

A geostationary satellite image

Because they pass frequently over the poles, these satellites provide more useful data at those latitudes than do the geostationary satellites. The infrared sensor on the geostationary weather satellites can distinguish areas as small as 4 kilometres in width, whereas the images in the visible light spectrum can resolve objects as small as 1 kilometre.

Satellite observations have an obvious advantage over areas of the earth for which there is little conventional weather data available. Tracking weather systems as they move across these areas has proven to be immeasurably valuable. Satellites are, in fact, used to produce "synthetic soundings" of the atmosphere over ocean areas where there are few aerological stations.

### Acoustic Sounders

These instruments, called SODARs (for sonic detection and ranging), use a pulse of sound to measure wind in the lower atmosphere. Their principal of operation is similar to that of radar except that they use sound waves rather than microwaves. When a sound pulse meets a discontinuity in the atmosphere, sound is scattered in all directions. The discontinuity can be in the temperature field, such as a strong temperature inversion. There can also be discontinuities in wind, associated with eddies that are present in the turbulent air flowing over the earth's surface. The SODARs' speaker directs a burst of sound, and a receiver then listens for a short period of time for the faint sounds that are scattered. The time from the initiation of the sound

173

burst to the receipt of the backscattered sound provides an estimate of the distance to the discontinuity, and the intensity of the return is a measure of the discontinuity's strength. The sounding systems can also detect slight changes in the frequency of the sound, and this Doppler shift is used to measure wind speed. Finally, the systems allow wind profiles to be obtained over the lower atmosphere to a maximum height of about 1.5 kilometres.

*Lightning Detection Networks*

Recently, lightning detection systems have been developed that use the electromagnetic wave properties of a lightning discharge to determine the location of the lightning stroke. As is readily apparent when listening to an AM radio during an electrical storm, flashes are accompanied by an immediate static burst on the radio. The strength of the noise (i.e., its amplitude) depends on the proximity of the radio receiver to the discharge. A network of detection stations has been established across North America, each with a detector that consists of an antenna that can determine the direction (or azimuth angle) of a lightning discharge from the antenna. Using the azimuth angles from several of the closest antennae, the discharge's location can be triangulated.

The system in use has a network with detector spacing of 200 to 400 kilometres, which gives it a data collection efficiency of up to 90 percent and a location accuracy of between 8 and 16 kilometres. Lightning strike information is analyzed by dedicated computing systems, and the location and density of discharges are displayed and updated every hour on a surface map. The movement of areas of lightning strikes and changes in the discharge rates can provide valuable information on the speed and rate of intensification of thunderstorms.

# Communication of Meteorological Data

Availability of appropriate weather data at a central location for analysis, forecast preparation and dissemination all depend on an accessible and reliable communication system. The invention of the telegraph in 1837 presented the first opportunity for meteorologists to undertake weather forecasting programs that could be useful to the public. Dedicated communications networks are used to send data to weather offices within minutes of the collection and analysis of the observation. The advent of the internet has made the communication task much less burdensome and has allowed the public to gain access to observations, analyzed data sets and all types of forecasts.

# Data Quality Control

Quality control of the observational data is essential to the weather forecast production system. Weather monitoring instruments are maintained and periodically calibrated to ensure that their operation is stable, and that measurement results are repeatable. At one time, trained weather observers and technical staff overviewed the data, but today computers do most of the work. Observations are stored in data banks, where quality control computer programs check for observational range, consistency with previous observations from the site and consistency with nearby observations.

# Chapter 10:
# Forecasting the Weather

Government organizations and the private sector both undertake weather forecasting programs. The government of Canada issues weather forecasts that "are provided for the good of the general population," according to federal policy. Governments worldwide provide a similar service for their populace. Provincial governments are responsible for forecast services that support sectors under their jurisdiction; the Alberta government operates weather forecast programs to support forest fire suppression as well as stream flow and flood warning services. Prediction services operated by the private sector serve the specialized needs

that fall outside the government's role. The main focus of this chapter is the federal forecasting process.

Weather forecasts are prepared and disseminated when they will be the most useful for clients. Warnings of hazardous weather conditions are prepared to enhance public safety and reduce property damage. Forecasts are also prepared to support many sectors of the economy, including transportation (aviation and marine transport), recreation (boating, skiing, avalanche warning) and construction. The public wants to know what weather conditions to expect for the day, and the weather forecast is a regular item on newscasts and in newspapers. In Canada, federal weather forecast centres are located across the country to partition the geographic area into manageable portions. These centres operate similar programs, which include some of the following services:

- preparing public forecasts and bulletins for cities

- issuing warnings for severe weather conditions

- preparing special sector-specific forecasts (transportation, recreation)

- providing information about unusual weather

- providing the latest weather conditions and historical climatic data.

### Prediction on the Large Scale— Guidance to the Forecaster

The weather forecast problem is complicated because the atmosphere extends well beyond the local forecasting area. Predictions of the atmosphere at the hemispheric and global scale are prepared at a central prediction centre and are sent to local weather offices. Meteorologists use this guidance material to prepare weather forecasts for several days into the future. In Canada, the central prediction facility is located in Montreal. Although this facility is mandated to support the national weather forecasting program, much of the meteorological information it prepares is also made available to provincial and private forecasting offices.

# Numerical Weather Prediction

The atmosphere is a continuous fluid that can be described using a series of mathematical equations that link temperature, pressure, wind, density and moisture. In 1904, the Norwegian meteorologist Vilhem Bjerknes recognized that these equations could be used to predict the future state of the atmosphere. In 1922, the British meteorologist L.F. Richardson first proposed a computational process to predict the weather using a simplified version of the theoretical equations. He proposed that a short-range forecast could be produced by several groups of individuals using arithmetic calculators to do the many necessary computations. One individual would coordinate the exercise, acting much like an orchestral conductor, ensuring the groups were working together to make the calculations faster than the actual weather was changing. Other individuals would take interim forecasts and make them available to users. Richardson envisioned having a warehouse to store the results for future examination and a research group to develop new forecasting techniques. He actually tried to prepare a six-hour forecast of pressure change using a version of this approach. After much laborious calculation, Richardson's prediction was off by a factor of about 100 because he did not appreciate some of the basic computational techniques that must be used in the process. The massive calculation exercise that was required and the resulting lack of success meant that further exploration of the technique was essentially abandoned for over a quarter of a

## Global Weather Prediction Grid

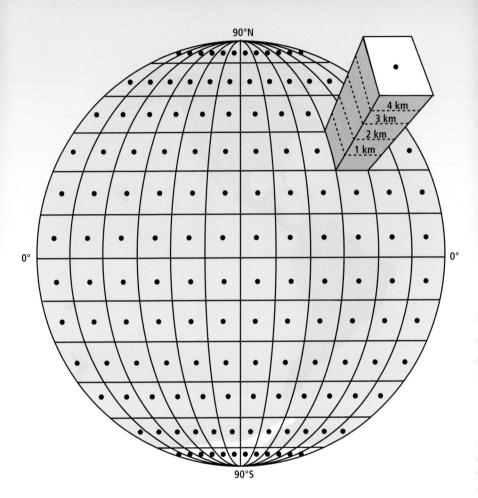

*Fig. 10-1* A coarse-mesh grid covering the earth's surface

century. However, Richardson's foresight is now considered quite remarkable.

The full set of equations needed to prepare a weather forecast is too complicated to be solved by simple mathematical techniques, and large computers must be used. The computer-based approach is known as numerical weather prediction (NWP). In the world of theoretical physics, the NWP approach is what is known as an initial-value problem: if the initial state of the atmosphere and the equations that govern its motion are known, its future state can be predicted. The equations must be in a form that allows observed data to be incorporated. This is accomplished through a

177

process called discretization. In the atmospheric NWP model, a three-dimensional grid of small cells is used to represent the atmosphere. Each grid cell has a horizontal length of a few tens of kilometres and a thickness ranging from tens of metres to a few kilometres. A network of several million grid cells is used in the largest and most sophisticated NWP models, which predict the atmosphere over the entire globe.

*Before modern-day weather forecasting, many people took their cues from nature. For example, when they saw ants moving to higher ground, people knew that the air pressure was dropping, so rain could be on the way.*

This three-dimensional grid needs to be initialized with data. Using data from observation stations, temperature, pressure, moisture and wind values are assigned to grid points located at each corner of each grid cell. At points above ground level, temperature, moisture, wind and pressure values are assigned using measurements from radiosonde balloon ascents and from instruments on-board aircraft. Synthetic data generated from weather satellites is used over areas largely devoid of observations, such as the oceans and polar regions. The rate of change of each of the atmospheric parameters is calculated using the equations of motion. At each grid point, a new value of the atmospheric parameter is calculated by multiplying the rate of change by a small time step that is a few

minutes long. A short-range forecast of all the parameters is produced, and from this a new rate of change is calculated and another short forecast is computed. This time-stepping process is repeated to generate forecasts out to a few days. The length of each time step is related to the size of the grid cells—the smaller the cells, the smaller the time step.

NWP was only made possible when large, numerical computers became available after WWII. The first successful experiments in NWP were completed at Princeton University in the late 1940s. Although it required almost 24 hours to prepare a one-day forecast using a simple model of the atmosphere, NWP's potential was demonstrated. Since the middle of the 20th century, most national weather predicting programs have used the NWP approach. The NWP computing centres use the fastest supercomputers available, and it still takes several hours to generate all the numerical forecasts. The NWP process is usually repeated twice daily, synchronized with the main 12-hour global weather observation cycle, and the guidance material is sent out to weather forecasting centres.

## The Forecast Products
### Warning Program

The cornerstone of the weather forecasting service is the warning program that issues messages alerting the public of potential weather hazards. Watches or warnings are issued depending on the expected severity of the anticipated event. A warning is issued when a weather event is expected to happen in a specific area, and a majority of people in the area may

## Numerical Weather Prediction Charts

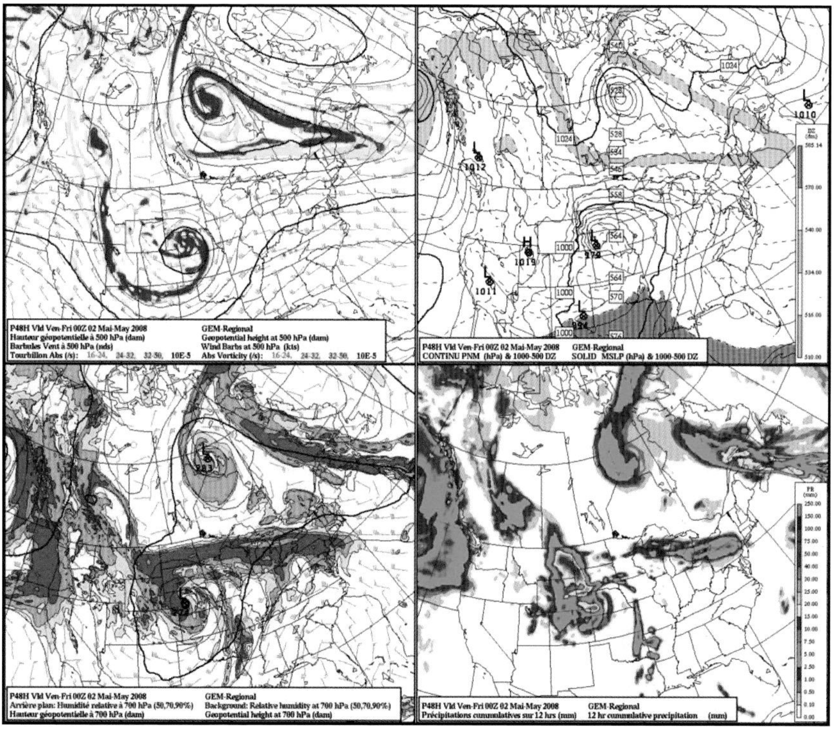

*Fig. 10-2* Upper left panel is the 500 mb forecast. Upper right panel is the surface pressure forecast. Lower left is the 700 mb forecast. Lower right is the precipitation forecast. All forecasts are 48 hours.

be affected. The rules under which a warning is issued are developed beforehand and are based on public safety considerations as well as the specific needs of weather sensitive sectors of the economy. Warnings are "short fuse" situations in which the atmospheric state is expected to change quickly. Severe thunderstorm warnings are issued two hours in advance of the event. Watches of severe, large scale events such as blizzards or wind storms are issued 12 hours in advance. The weather watch is a "heads up" that is issued when the forecaster expects that a severe weather event may occur in the near future. Watches

are issued up to six hours in advance of a severe summer storm, and up to 24 hours in advance of a winter storm. The forecaster also prepares special weather statements to provide additional explanations of weather conditions that will have a public impact but are not expected to reach the warning or watch criteria. (See "Three Types of Weather Alerts" chart on p. 144 for more information.)

### Aviation

The aviation industry requires detailed forecasts of many atmospheric parameters, and several different products are prepared. The Terminal Aerodrome

Forecast (TAF) is used to enhance aircraft landing and takeoff safety at airports. The TAF includes wind speed and direction, precipitation type and intensity, type of visibility obscurations (fog, precipitation, smoke and dust) and the height of cloud above ground level at the airport. The forecasts are issued every six hours, out to 12 or 18 hours in the future. They are specific about the times that different weather conditions can be expected to change.

Forecasts are also prepared in a graphical form for the area in which aircraft are flying. These maps depict inflight weather conditions across the area and are used by aircraft pilots to infer weather conditions at airports where no TAF is issued. These Graphical Forecast Area charts are issued every six hours for six hour intervals, out to 18 hours. The maps include the pressure pattern and frontal position forecasts, and they depict areas of cloud, including the expected height of cloud bases and tops, and areas where cloud bases will be near ground level. Areas of precipitation and regions where visibility near the ground might be obscured are also shown. Identifying regions in which aircraft icing conditions might be expected is important for safe aircraft operations, so these areas, including anticipated icing rates, are depicted. Atmospheric turbulence is a concern for passenger comfort and aircraft safety, so areas of turbulence including the degree of severity and levels in the atmosphere where turbulence can be expected are shown. The wind pattern at flight level is useful for aircraft navigation and planning. Maps of winds at several levels in the troposphere are generated from the numerical weather prediction models that provide guidance to the weather forecaster. Again, these products are issued for six hour intervals, out to 24 hours.

## Summer Severe Weather

The summer severe weather program usually focuses on convective weather and severe thunderstorms. The period of concern extends from late spring until the end of summer. When considering a convective weather forecast, the meteorologist is most concerned about the vertical thermal stability of the atmosphere. The development and movement of convective clouds is the forecasting issue. Where will they form, and will they grow explosively to produce heavy rain, hail, wind storms or even tornadoes? The severe weather forecaster needs to have a good three-dimensional picture of the atmosphere. This picture is built up using radiosonde observations, as well as analyses and predictions of temperature, moisture and wind patterns from the ground up to the tropopause. Images from weather satellites and weather radar provide valuable detail about the structure and motion of the weather systems and existing convective storms. The forecaster must be attuned to the influence of low level moisture from sources such as transpiring forests or croplands. Thunderstorms can develop and grow to severe intensity over a period of an hour or so, and they may move rapidly. Watches and warnings for specific sites need to be issued promptly for areas where severe thunderstorms are anticipated so that the public can be warned.

Particularly challenging are forecasts of tornadoes, which typically have a lifespan of a few minutes to one hour. Every severe thunderstorm is not a tornado producer. Even sophisticated observational tools, such as the weather radar, do not have high enough resolution to see the small tornado funnel. However, radar can measure wind speed over larger scales using what is known as the Doppler shift technique. Because tornadic thunderstorm cells are usually rotating, the Doppler radar can frequently detect the large-scale rotation of a supercell thunderstorm. The regular observation network is too sparse to delineate small-scale features or observe small-scale events, so the severe weather forecaster depends on observations made by a network of voluntary severe weather watchers. Only when a tornado has been observed to touch the ground is a tornado warning issued. Time is of the essence in issuing the forecast and getting the warning to the public. When the warning is issued by the weather office, the Alberta Emergency Public Warning System is activated. This system was established after the Edmonton tornado of 1987 to warn the public of disastrous events.

> A lot of people like snow. I find it to be an unnecessary freezing of water.
>
> –Carl Reiner

## Winter Severe Weather

Winter severe weather includes heavy snowfall or rainfall, freezing precipitation, wind chill and strong winds. In Alberta, many of these conditions can occur during any season. Winter severe weather usually takes a longer period to develop than does severe summer weather. The large-scale prediction of weather patterns is heavily relied on to determine the location, intensity and motion of the winter storm event. The NWP products provide good guidance, but errors in the position of the features can make all the difference in forecasting precipitation rates or blizzard conditions, and careful attention is paid to pressure patterns and their evolution.

Weather observations play a key role and, observations from volunteers are relied on to fill in the details. The start of snowfall is monitored at observation stations, and the forecaster does subjective correlation between snowfall observations and the precipitation patterns displayed by weather radars.

Temperature and wind predictions are used to prepare wind chill forecasts. Areas of above-freezing temperature layers above the ground determine where freezing precipitation may occur. Freezing precipitation is usually difficult to forecast more than a few hours in advance—a problem that is similar to the summer severe weather forecast. There is a need to anticipate the area where the event may happen and then use all observational data for clues to its occurrence and intensity.

## Marine Forecasts

These forecasts are issued for large inland lakes during the marine shipping season. In Alberta, the marine forecast is only issued for Lake Athabasca. For smaller waterbodies, users are expected to use the public forecast and pay close attention to the relevant forecast parameters. For mariners, the most important parameter is the wind, which can be derived accurately from maps of surface pressure. Waves are built up from the wind blowing for a long time across a water body. Wave height is linked to wind speed, the duration of the wind and what is known as the fetch, which is the length of water over which the wind has blown. Freezing spray is also a concern when air temperatures are expected to be below 0° C over fresh water surfaces. Because metallic ship superstructures cool to the air temperature, the spray from waves breaking over the ship freezes on contact, and ice can build rapidly. The forecasts issued in late autumn during periods of cold outbreaks include a freezing spray warning. Horizontal visibility is also a concern for safe marine navigation. Fog is the main culprit that affects visibility in marine forecasts. The lake acts as a ready source of moisture, and it can also cool air passing over the surface to below the saturation point.

## The Role of the Forecaster

Meteorologists in the weather centre in Edmonton prepare all public and marine forecasts and warnings for Alberta. They also produce the aviation forecasts for western Canada. Six to eight forecasters are on duty every shift, and the team works together to make consistent decisions on the evolution of weather systems across the forecast region. The forecasters in a weather centre talk to their counterparts in the offices in adjacent regions to ensure that the weather systems are handled consistently as they move across the country.

The weather forecast program operates 24 hours per day every day of the year, so shift work is the reality for forecasting staff. The meteorologist plays a vital role in the weather forecasting process. Although observational technology and computerized forecasting tools are sophisticated, the weather forecasting process requires a human touch. The quick synthesis of data, the ability to recognize patterns and the ability to make decisions in the face of conflicting or incomplete data are tasks best left to the human forecaster.

The first task for a meteorologist starting a shift is to get a briefing from staff members who are finishing a shift. Which weather systems are active in which sections of the forecast region? What forecasts and warnings have been issued? What is the confidence level that conditions will evolve as predicted? Today's weather office is moving toward a paperless working environment, and the computer-supported graphical workstation has replaced the map displays that were once the predominant feature. The meteorologist reviews observed conditions over the region of forecast responsibility and often undertakes a hand analysis of the computer-plotted surface weather map. This is one of the few paper products still used because the analysis process helps focus the meteorologist's attention on the details and occasional data anomalies and inconsistencies. Hourly weather data from automated and staffed observation sites across the forecast region is continuously streaming into the office through the communications system. Computers organize the data, and computer algorithms compare the observations with the valid forecast. If there is a significant discrepancy, the meteorologist is notified so that an amendment can be issued.

A large variety of other depictions of the atmosphere are generated. Weather patterns at various levels in the vertical, and graphs of the vertical temperature and moisture structure provide the meteorologist with knowledge of atmospheric conditions favourable for convection. Imagery from geostationary and polar-orbiting satellites is continuously displayed. Displays of weather radar images provide a time sequence of cloud patterns and precipitation rates across the province, and the lightning detection system indicates where thunderstorms are active. The meteorologist uses his or her knowledge of atmospheric processes and the influence of topographic and other controlling features to diagnose the current atmospheric condition and its short-term change.

> **Old weather saying:**
> *As high as the weeds grow,*
> *so will the bank of snow.*

The forecast products are all prepared at the computer workstation. The computer automatically generates a time sequence of all the weather elements, such as cloud, temperature, wind, precipitation and visibility, to be included in the forecast for all locations in the region. The meteorologist makes any necessary adjustments to the weather element sequence using an interactive knowledge-based system. When the forecaster is satisfied with the forecast, a command is sent to the computer system to generate the worded forecast, which is transmitted to media outlets. The product is also sent to a voice generation system that automatically prepares an audio clip for broadcast on the weather radio network across the province.

At the computer screens prior to the April 20, 2008 spring snowstorm.

For aviation, Terminal Aerodrome Forecasts are prepared for each of the larger airports across the province. Again, the computer guidance material is used but the worded forecast is input directly into a bulletin by the meteorologist. Aviation forecasts include many weather parameters, and a continuous comparison is made between observations and forecasts. The comparison is largely performed by the computing system, which automatically sends out a warning to the forecaster if discrepancies between the forecast and observed values exceed predefined limits. The meteorologist must quickly evaluate the reasons for the change and issue an amendment to the TAF.

The forecaster also composes prognostic Graphical Forecast Area charts directly on an interactive aviation workstation. Again, the material generated by the central computer is used as a guide. The forecast surface pressure patterns are based on the NWP guidance. The forecaster adds frontal positions and outlines areas of cloud and precipitation, as well as regions where icing and turbulence are to be expected. All these fields are developed from the temperature, moisture and wind patterns in the NWP guidance.

The Graphical Forecast Area chart

Weather forecasting screens at the Edmonton weather office.

# Chapter Eleven:
# Weather Modification in Alberta

There has been a long-standing interest in modifying the weather, which belies the saying "everyone complains about the weather, but no one ever does anything about it." Weather modification has, in many cases, been inadvertent, such as the cooling caused by massive injections of volcanic dust or plumes of smoke from fires. Usually, the impetus for deliberate weather modification has been precipitation—either to stimulate the atmosphere to produce more, or to change the type and amount that is falling. During the droughts of the 19th and 20th centuries, many individuals crossed North America offering their rainmaking services to the beleaguered farming community, with mixed success. When there are few or no clouds (such as during a drought), no amount of artificial weather modification can produce rain.

> **Sunshine is delicious, rain is refreshing, wind braces us up, snow is exhilarating; there is really no such thing as bad weather, only different kinds of good weather.**
>
> —John Rushkin

Artillery used to deliver silver iodide to storms in China

freezing conditions overnight. Many citrus fields in Florida have large fans to maintain turbulent flow near the ground, inhibiting below-freezing temperatures. Crops are also coated with water so the latent heat released when the water freezes keeps the temperature of the vegetables above freezing. And, of course, it is common practice to cover tomatoes and other susceptible plants to ward off early-fall frosts.

Weather modification has been a focus for international scientific research and has attracted the attention of government policy makers. Intensive scientific investigation of the potential for weather modification has been underway since the middle of the 20th century. Experimental work got underway in the late 1940s with Project Cirrus in the U.S., which was initiated to investigate the possibilities and limitations of weather modification. Today's precipitation-oriented weather modification approaches use techniques based on our understanding of precipitation formation. As discussed in the chapter on precipitation, the Bergeron process proposes that in clouds containing a mixture of ice crystals and supercooled water droplets, ice crystals rapidly grow at the expense of droplets. The process continues until the ice crystals grow large enough to start falling as precipitation. In the late 1940s, researchers at the GE Research Laboratory in Schenectady, New York, were actively exploring cloud modification sciences. In 1946, Vincent

Weather modification ideas were first centred on brute force approaches. Proposals were made to divert airflow from moist to dry areas, and to stimulate the formation of cloud by initiating updrafts. One of the more popular approaches was to use sound to modify rainfall. In England, people claimed that the sound of ringing church bells somehow reduced the severity of rain. The practice of ringing them during thunderstorms was banned, however, because of the high fatality rate of bell ringers being struck by lightning in their bell towers. Many approaches to enhance rainfall also used loud noise. Producing shock waves by firing upward-pointing cannon or by detonating balloon-borne explosives were popular, but unproductive, techniques.

A more successful brute force approach is still used on a small scale to modify temperatures and thwart the onset of

Shaefer discovered, in what he termed a "serendipitous occurrence," that dry ice could be used to cool cloud droplets to the freezing point, producing ice crystals. Later that year, Bernard Vonnegut found that tiny crystals of silver iodide behaved like ice crystals and acted as growth nuclei. Of particular importance was the observation that silver iodide also became active growth nuclei in clouds that had few natural ice crystals present because of the temperature. Both the dry ice and silver iodide crystal processes are effective in producing huge numbers of ice crystals. This stimulation of ice crystal formation has been used to break up decks of stratus clouds, to dissipate supercooled fog decks at airports and to modify rainfall and hail production processes.

A hail cannon produces a shockwave to stimulate rain.

> Whether the weather be hot,
> Or whether the weather be
> not, we'll weather the weather,
> whatever the weather,
> whether we like it or not!
>
> —Unknown

During the 1950s, experimentation centred on means to deliver finely ground dry ice and silver iodide into appropriate precipitation growth regions of clouds. Dry ice, or solidified carbon dioxide, is relatively inexpensive and, when ground into a fine powder, can be readily dispersed by aircraft. Silver iodide crystals are produced in large quantities when flares containing the chemical are burned. Attempts to burn flares at ground level, with the expectation that the crystals would be carried aloft by the turbulent wind into the cloud base, have proven to be inefficient. Small rockets and artillery shells containing silver iodide have also been used to deliver the chemicals to the cloud. All the ground-based techniques have the inherent problem of trying to deliver crystals into clouds far removed from a point on the ground. A more effective approach uses aircraft to deliver the silver iodide crystals directly to the desired cloud regions. Flares can be dropped into the selected clouds, or wing-mounted flares can be ignited when the planes fly into the cloud region where precipitation modification is desired.

An intense storm moves into the Calmar area south of Edmonton in October of 2002.

## Legal and Equity Considerations

The potential for weather modification to have non-domestic applications has been a concern in many political jurisdictions. At the international level, the United Nations passed a resolution in 1978 that prohibits the use of weather modification techniques for military applications. The Canadian Weather Modification Act does not prohibit weather modification but requires that notification of intent to conduct operations and a record of the activity be provided to Environment Canada. Similarly, Alberta uses a non-obtrusive approach; under the Municipal Government Act, municipalities wishing to undertake weather modification programs must conduct a plebiscite in the affected populated area.

Equity issues have also been raised with respect to the enhancement of rainfall by cloud seeding. The complexities of precipitation formation processes and the assumption that an area can be targeted for enhanced rainfall is questionable; that is, if there is only a fixed amount of moisture available for rain enhancement, does stimulation in one area lessen rainfall in areas farther downwind? A similar issue comes up when water needs to be allocated from streams in semi-arid areas, such as

*It takes the same amount of precipitation to make 2.5 centimetres of rain as it does to make about 25 centimetres of snow.*

southern parts of Alberta and Saskatchewan. Interprovincial water allocation boards are in place to ensure that scarce water resources are shared along river courses. With a rainfall enhancement process, however, one can assume that the moisture is only temporarily extracted from the clouds in one area and deposited on the ground. Much of the deposited moisture subsequently evaporates and is then converted naturally by the atmosphere into rain clouds downwind. With this view, the stimulation of rainfall in one area enhances the moisture regime over a larger region.

## Hail Suppression in Alberta

Hailstorms have a long history of causing damage to crops, especially in south and central Alberta. A series of damaging hailstorms in the early 1950s prompted the farming community to push the provincial government for action. The early success of experiments in the United States appeared to hold a potential solution. The Alberta Research Council was charged with taking the lead and, together with the federal weather service and the McGill University Stormy Weather Group, initiated a program to examine the issue. Because weather modification science was still in its infancy, it was not considered feasible to immediately embark on a weather modification program. An intensive study program was launched in the mid-1950s to examine hailstorm climatology, to understand hailstorm characteristics and, in particular, to understand the hail-forming process of Alberta thunderstorms. The studies were completed

throughout the 1960s, and weather modification experiments undertaken elsewhere had apparently yielded positive results. In 1969, a cloud seeding program based out of Red Deer was initiated over central Alberta.

Silver iodide flares mounted on aircraft wing

The conceptual approach was to stimulate production of large numbers of graupel that would then compete with natural graupel particles for available liquid water in the cloud system. It was further assumed that cloud seeding would not change the energy of the thunderstorm. Although it was possible that an energetic storm might increase the total available moisture by increasing the strength of the inflow, the competitive growth concept worked on the assumption this would not happen. Instead, the concept assumed there would be a shift of the hail size distribution, from a small number of large, damaging stones to a larger number of small stones that would cause less damage. Hail-producing thunderstorms are complex and have a multi-cellular structure consisting of the large main cumulonimbus cloud and numerous feeder clouds, which move into and become part of the main storm. The hailstorm model assumes that the hail embryos are produced

191

## Competitive Growth Concept for Hailstorm Seeding

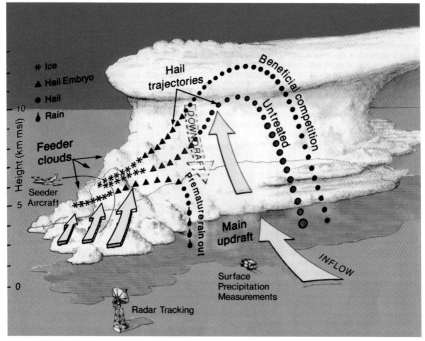

*Fig. 11-1*

primarily within the feeder clouds, and enhanced production of embryos leads to competitive growth of hailstones. Several aircraft carried silver iodide flares to seed the feeder clouds, initiating the competitive growth process.

As well as the hail reduction program, an experimental rainmaking program was also conducted using the silver iodide seeding approach. The objective here was to enhance rainfall from cumulus clouds that would not necessarily evolve into rain-producing clouds. Some experiments using ground-based seeding generators were also undertaken, but this method was ineffective in getting seeding material into target clouds.

The Alberta Hail Project (AHP) lasted for three decades and had a number of benefits. It demonstrated that rainfall could be stimulated in some clouds and enhanced in hailstorm feeder clouds. The seeding process occasionally caused hail embryos to fall out of the clouds early. Although the AHP was unable to demonstrate conclusively that more hail embryos resulted in smaller hail on the ground, the program had many outstanding scientific achievements in severe weather research. It made innovative use of weather radars that allowed detailed examination of thunderstorms, which gave new insights into the structure of storms and provided the ability to detect regions within clouds where hail had formed. As well, the program did much to enhance knowledge of hailstorm processes, their link to larger atmospheric flow, and the roles that the high plains and mountains play in thunderstorms. The AHP also developed practical systems that have been used around the world for seeding multi-cell hailstorms. The program led to improvements in hail and tornado forecasting procedures used to this day.

## Insurance Industry and Hail Suppression

Although the Alberta Hail Project ended in 1985, it was followed in 1996 by a hail suppression program focused on reducing hailstorm damage to property. The program was initiated after a severe hailstorm struck Calgary in 1991, resulting in $300 million of insured losses. The Alberta Hail Suppression Project (AHSP) was the first to be funded privately by insurance companies rather than by the government. A preliminary five-year seeding program operated from 1996 to 2000, which indicated that insurance hail losses were about 50 percent less than expected. Because of its success, the program was officially implemented and is still running today.

The AHSP follows the same scientific approach used by the AHP and assumes that the competitive growth conceptual model is a valid approach to reduce the size of hailstones. Operations are centred on the hail-prone corridor between Calgary and Red Deer, and it operates seven days per week during the June 1 to September 15 hailstorm period. Candidate clouds that pose a potential hail threat to urban areas are identified using a weather radar located at the

The weather radar at Olds-Didsbury airport with a candidate hailstorm in the background.

Olds-Didsbury Airport. Three seeding aircraft deploy silver iodide nuclei into the top and base of clouds.

The program's success is obvious when trends in insurance claims in hail-prone areas of Alberta are compared with those from elsewhere in the world. Hail claim trends are reduced in the seeded area of Alberta relative to trends in non-seeded areas.

*During World War II, gas was poured into pipes or trenches on each side of some fog-bound runways and set on fire. The flames provided enough convection to lift the fog so that planes could land. Hundreds of thousands of litres of fuel were burned for this purpose, with as much as 380,000 litres being used per hour.*

193

# Chapter Twelve: Air Quality

Alberta has a diversified economy and a growing population base. Although population growth in Alberta's cities is more rapid than in the rural areas, air pollution sources span the province.

## Emissions to the Atmosphere

Activity associated with oil and gas exploration, production and processing is the largest source of atmospheric emissions. Agriculture, home heating, transportation and electric power generation are also major emission sources. Pollutants are classed as either primary or secondary. Primary pollutants have well defined sources such as smokestacks and vehicle tailpipes. Secondary pollutants are those that are the product of chemical reactions in the atmosphere. These reactions occur among the primary pollutants and with various constituent gases in the clean atmosphere.

The high temperature combustion of fuels produces most of the emissions of concern. Nitrogen oxides are produced when any combustion occurs in the oxygen and nitrogen gas mix of the atmosphere. They may also be produced by the oxidation of nitrogenous compounds that are in the fuel. Nitric oxide (NO) and nitrogen dioxide ($NO_2$)—collectively known as $NO_x$—and nitrous oxide ($N_2O$) are the most important nitrogen pollutants. Nitrous oxide is also significant because of its radiative properties. It strongly absorbs infrared radiation and is one of the main greenhouse gases. Stationary sources (power plants and residential and industrial heating) and mobile sources (vehicles and aircraft) contribute, and Alberta accounts for about 33 percent of Canada's $NO_x$ emissions. Carbon monoxide and carbon dioxide are also products of combustion and are formed when the carbon in fossil fuels is oxidized by atmospheric oxygen. Many fossil fuels contain trace amounts of sulphur and, though much of the sulphur is extracted from fuels, residual amounts are converted to sulphur dioxide ($SO_2$) during combustion. Hydrogen sulphide ($H_2S$) is naturally present in natural gas and oil reservoirs. It is a toxic gas with a characteristic rotten egg smell. In the production of fossil fuels, $H_2S$ is extracted and converted to elemental sulphur, but small amounts of residual $H_2S$ are released or transformed into $SO_2$ through combustion. Alberta's $SO_2$ emissions are about 22 percent of the Canadian total.

A wide variety of hydrocarbons find their way into the atmosphere. These volatile organic compounds (VOCs) originate from the evaporation and combustion of liquid fossil fuels and from chemical manufacturing, petroleum production and solvent use in chemical processes and domestic applications. Alberta contributes about 26 percent of the Canadian VOC emissions.

Ammonia ($NH_3$) plays a role in the production of particulate matter (PM) in the atmosphere. Animal husbandry, fertilizer application and some industrial processes are sources of ammonia in Alberta.

Atmospheric particulates are contaminants that are attracting increasing concern. Some particulates are classed as primary pollutants, but many particulates are secondary pollutants that form in the atmosphere. Particulate matter impairs the clarity of the atmosphere by degrading its light transmission properties, resulting in reduced visibility of distant objects. It can also have negative health impacts. Particulates are generally categorized according to their aerodynamic size based on their ability to be deposited in lungs. Larger particles usually have a natural origin, such as wind-blown dust from dry soil or re-suspended dust from road surfaces, but can also originate from combustion as smoke, soot and ash. These particles are generally larger than 10 microns in diameter.

It is the smaller particles—generally with a size less than about 2.5 microns—that find their way into the lungs and cause the greatest health risk. These particles may be either solid or liquid droplets and are often formed from the gases that are a product of combustion.

Other compounds that are in the categories of toxics, heavy metals and persistent organic pollutants are either minor by-products of industrial processes or are heavily regulated, and few reach the atmosphere in Alberta.

Emissions of pollutants vary greatly across Alberta. With their high population concentrations and intense levels of industrial activity, urban centres and their surrounding areas emit the majority of pollutants. The oil sands region has many emission sources, and the half-dozen coal-fired power plants across the province are also heavy polluters. Agricultural emissions are widespread in the cultivated lands in the southeast quarter of the province, but higher emission rates are associated with the feedlot operations in southern Alberta. Because gas and oil extraction and processing facilities are spread across the province, pollutants from these sources have a wide geographic distribution. Industries report their annual emissions to the government of Canada for archiving in the National Pollution Release Inventory. Alberta has more than 9000 point sources that are in the inventory—more than twice as many as in any other Canadian jurisdiction.

There is an atmospheric cycle through which pollutants are mixed, moved around, sometimes transformed and

finally removed from the atmosphere by several natural processes. Polluting gases are diluted downwind of the emission sources. Although the saying "the solution to pollution is dilution" was often heard, dilution is taxing on the clean air resource. Consider vehicle emissions of carbon monoxide. In 1965, the average emission rate was 30 grams of CO per kilometre of travel. About 2 million litres of clean air—the amount breathed daily by 200 people—are required to dilute that much CO to levels that are safe for humans to breathe. So, it takes large volumes of air to dilute all the pollutants to levels that are considered safe. In many regions of the world, there is not enough clean air, and an emphasis was placed on improving vehicle efficiency along with reducing pollutant emission rates. By 1997, the emission standard for CO was reduced to 2.1 grams per kilometre. On a per vehicle basis, we are now polluting significantly less air.

Cloud condensation nuclei are in part composed of the small particles originating from pollutants. Studies of rainfall patterns have also revealed how pollution in the atmosphere influences precipitation. For example, enhanced rainfall within Chicago relative to its less polluted suburbs has been attributed to the vast number of condensation nuclei injected into the atmosphere by industrialized regions in that city. Farther downwind, enhanced rainfall in the city of LaPorte, Indiana, has been attributed to the transport of nuclei from the same upwind Chicago pollution sources. The rainfall process is also a cleansing mechanism for polluted atmospheres—it removes and washes away some nuclei. In anticipation of the 2008 Olympics, China has suggested it will artificially enhance precipitation in Beijing to rain out the heavy pollution concentrations and improve air quality.

## Pollution Movement in the Planetary Boundary Layer

The lowest layer of the atmosphere is known as the planetary boundary layer (PBL). Within the PBL, most life forms are subjected to air pollution. Air quality and the dispersion of pollutants are influenced by the turbulent motion and temperature structure within the PBL. Three key factors determine the characteristics of the PBL:

i) The transport of heat from the earth's surface. Heating caused by the solar radiation cycle, heating from phase changes of water (latent heat) and the spatial variation in heating (near lakes and over different land surfaces) are all examples of the complex heating process. Heat is an important energy source that leads to mixing of the PBL—just like the heating element on a stove causes water in a pot to turn over and mix.

ii) The influence of roughness at the earth's surface. Airflow is disturbed by the presence of objects, from blades of grass or trees to buildings, hills and mountains. These objects result in a surface roughness that slows the airflow in the PBL by imposing what is known as frictional drag. Large objects can also deflect and block airflow. Roughness features induce mechanical turbulence, much like water flowing over a rough riverbed is turbulent. The larger the roughness features and the stronger the speed of the flow, the greater the degree of turbulence.

iii) Airflow and thermal structure of the atmosphere at the top of the PBL. This level is the bottom of what is known as the free atmosphere and is characterized by the atmospheric flow associated with large-scale pressure patterns. At this level, the geostrophic wind law describes the horizontal flow. Upward and downward motions associated with the large-scale flow are also evident. The top of the PBL can be a sharp transition zone into the free atmosphere. If the temperature in the PBL is lower than in the immediately adjacent free atmosphere, the top of the PBL effectively acts as a lid that can prevent the upward motion of pollutants out of the PBL.

## Turbulence

Atmospheric motion is anything but uniform—it is turbulent. Whether winds are light or strong, we sense the turbulence as gustiness. When wind measurements are recorded, the time series shows frequent changes in wind speed and direction. These fluctuations are characteristic of this turbulent flow. Air quality scientists think of wind as having two components. The first component we can call the wind flow. It is considered to be steady for some interval of time. The second component is the perturbation, which is the slight departure from the mean wind. The motion of air pollutants is governed by the wind flow itself and by perturbations in that flow. Wind flow is a vector quantity—it has speed and direction, and it is governed by the mechanical equations of motion of the atmospheric fluid. The mathematics dealing with perturbations in the wind, known as the turbulent component of the airflow, are complex. An old saying that demonstrates the difficulty of understanding turbulence goes "when you get to heaven, God will offer two explanations—one for quantum physics, and the second for turbulence theory." Air quality science focuses on the effects of such turbulent motion.

The rate at which pollutants disperse and how pollution concentrations change with time is described by the mathematics of turbulence theory. This theory can be put in the form of mathematical equations, which are used to calculate the rate of spread of pollutants and the changes in pollutant concentration downwind of an emission source.

When pollutants are emitted into the atmosphere from a point source, such as a smokestack, they have two key characteristics. If they are warmer than the surrounding air, their buoyancy causes the pollutants to rise. They also immediately start mixing because of turbulence and spread out laterally from the source. If there is a wind associated with the fluid, the pollutant cloud takes the form of a plume. The plume has a characteristic shape, as shown in Figure 12-1. It initially moves vertically then is bent over by the wind. With time, the polluting gas spreads away from the centre line, taking on a conical form. The shape of the cone, and the pollutant concentration within the cone, is described by the statistical properties of the turbulent atmosphere.

## Temperature Structure

The temperature structure of the planetary boundary layer has an important influence on air quality because the vertical temperature structure controls the vertical mixing. As mentioned earlier, the rate of change of temperature with height is known as the lapse rate. The lapse rate is usually negative—that is, temperatures decrease with increasing height. The greater the rate of temperature decrease with height, the greater the ability of the atmosphere to disperse pollutants. Most pollutants are emitted near the ground, so the lapse rate in the PBL is important in dispersing pollutants. However, there are many situations in which temperatures increase with height, creating an inversion. An inversion condition in the atmosphere slows the dispersion

## Depiction of a Plume Rising in a Stable Atmosphere

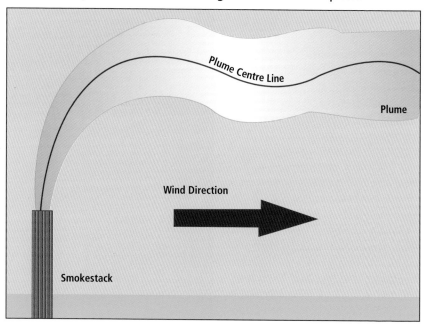

Fig. 12-1

process, and pollutant concentrations can continue to increase near a source that is continuously emitting.

The lapse rate varies widely during a 24-hour period. When the sun sets, the ground surface cools, as does the air in contact with the ground. Air farther aloft does not cool as quickly, and a nocturnal inversion develops. Pollutants emitted into this inversion layer may move laterally with the wind, but they have little vertical movement. The nocturnal inversion often traps pollutants near the ground.

Other types of inversions are also important in Alberta. The interface between two different air masses is known as a front. Frontal inversions can play a role in trapping pollutants near the ground. The frontal surface can have a shallow slope, and a large area of the province can be subjected to frontal inversion effects, especially in the cold seasons with slowly moving warm fronts. Cold fronts, on the other hand, are often rapidly moving, and they clean out stagnant polluted air masses.

Terrain can also play a major role. Cooler air pools in river valleys and may be overrun by warmer air. The resulting inversion can trap pollutants within a valley. The Rocky Mountains play a more significant role. In the lee of the Rockies, winds with an easterly component may push colder air up against the mountains. Often, cold arctic air at ground level is overrun by warmer Pacific air. This damming effect can result in a prolonged period of inversion in the PBL in the lee of the Rockies. Near the mountains in winter, such a situation usually

ends with the onset of a Chinook when strong, turbulent westerly winds replace the light easterly flow.

In December 1952, more than 4000 people in London, England, died as a result of a severe pollution event. A strong thermal inversion acted as a lid and trapped moisture and pollutants in the PBL above the city. Incident sunlight was reflected by higher level cloud decks and by the top of the fog layer itself. Sulphur-laden smoke from coal-fired furnaces and stoves accumulated over a period of five days as the inversion persisted. Most people died because of respiratory tract infections, such as bronchitis and bronchopneumonia. Largely because of this event, the term smog—meaning a mixture of smoke and fog—was coined. Alberta never experiences such sulphurous smog events, but emissions of sulphur gas (usually $SO_2$) have been the main focus of air quality regulation and control in our province and elsewhere in the world.

*London, England, used to be known for its so-called peasoupers: heavy blankets of smog, made up of a combination of fog and smoke from factory smokestacks and the burning of coal. In December 1952, 4000 people died of chest disease as a result of pea-soupers.*

## Formation of Photochemical Smog

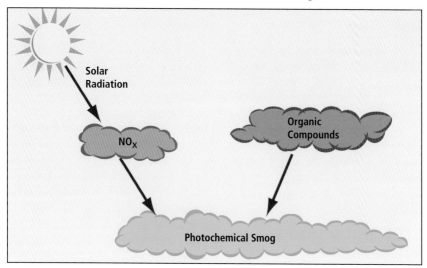

*Fig. 12-2*

## Chemical Transformation in the Atmosphere

Photochemical smog is common in all large cities where vehicles, industries and other pollution sources are concentrated. Most prevalent and well known were the smog events that frequented greater Los Angeles. Through the 1960s and '70s, the causative chemicals and chemical reactions responsible for the smog were determined, and a solution to the issue has progressed with California's imposition of stringent vehicle emission controls. Photochemical smog is now a much less frequent occurrence, and events are less severe than during its peak in the 1970s.

Photochemical smog consists of a mixture of the primary emissions of $NO_X$ and VOC gases, and the secondary pollutants ozone ($O_3$) and particulates. These two secondary pollutants are produced when intense sunlight acts on the mixture of primary emission gases. Reactions in the polluted mixture proceed depending on the concentrations of the precursor gases, the temperature and the amount of mixing that is present in the atmospheric boundary layer. Some of the reactions are rapid, some are slow and some depend on the intensity of sunlight. Generally, the reactions in the polluted soup proceed to what is known as an equilibrium state. In the worst-case scenario, this equilibrium state has high concentrations of ozone and particulates.

Alberta's major cities have significant emissions of the pollutants that can lead to the production of photochemical smog. The presence of $NO_2$ is evident because this gas preferentially absorbs light in the short wavelengths (blue to green). When light passes through the gas, these short wavelengths are absorbed, allowing the longer wavelengths—the red to orange colours—to pass through the gas. As a result, the gas appears to have a brownish colour. The brownish tinge is usually evident downwind of Edmonton and Calgary during light wind conditions in summer and winter. Both cities can have meteorological conditions in summer that favour production of photochemical smog (calm winds and intense sunlight). Because the reactions can take a couple of hours, peaks in concentrations of ozone and small particulates vary in space and with time. Air quality monitors near downtown Edmonton and Calgary usually show elevated $NO_X$ concentrations, especially during the morning rush hour when there is more $NO_X$ available than is necessary for ozone production. During a typical day, $NO_X$ levels peak during morning and afternoon rush hours. Ozone reaches a peak late in the afternoon when solar intensity reaches a peak downwind of the city centre and the photochemical conversion process is most active. Overnight, concentration levels of most of the primary and secondary gases drop to a minimum.

> **The Sun himself is weak when he first rises, and gathers strength and courage as the day gets on.**
> —Charles Dickens,
> *The Old Curiosity Shop*

Smog situations can also occur in winter. They characteristically have high concentrations of PM and $NO_X$, but ozone levels are low because of low temperatures and weakened solar intensity. Winter smog usually forms because of the presence of strong thermal inversions and low wind speeds that trap pollutants near the ground. Chemical reactions proceed slowly at low temperatures, though low temperature conditions are favourable for the formation of some types of particulates. Stagnant pollution trapping events are more common in late autumn in Edmonton and Calgary.

The formation of acids in the atmosphere and the deposition of these acids on lakes and forests is an issue in much of eastern North America and Europe. Sulphur dioxide and $NO_X$ both play a role, and the high rates of emissions of these gases in Alberta have been the focus for a considerable research effort. Sulphur dioxide has been the primary culprit in the acid rain issue. It transforms into sulphuric acid in the presence of water vapour and other atmospheric chemicals known as free radicals. Nitric acid ($HNO_3$) forms in the atmosphere as a byproduct of the same chemical reactions involving NO and $NO_2$ that are responsible for ozone production. The gaseous forms of the acids can be deposited in dry form, whereas the acids in aqueous form are deposited in precipitation known as acid rain.

Acid-rain gauge with cover

An aircraft and ground vehicle used in an air quality research study during 2005

Negative impacts of acid deposition usually depend on the nature of the receptor. Acids can be neutralized or rendered less harmful by some chemicals that occur naturally in many receptors. Most of the receptor surfaces in Alberta have this neutralizing capability, known as a high buffering capacity. However, the groundcover and lakes in areas of Precambrian rock (the shield areas of northeastern Alberta) have low buffering capacity and can be susceptible to acidification. As well, because the acids and the chemicals that are acid precursors can be transported long distances by the winds, acid deposition can occur at far from the pollution source. This is the reason for much of the acidification of sensitive lakes in eastern Canada and the United States. Many of the polluting sources are well upwind, and long-range transport of sulphur compounds is the primary culprit. In the case of Alberta, considerable attention has been given to the potential for pollutants in Alberta to contribute to acidification in Saskatchewan.

Forest fires can also have a major influence on air quality in Alberta. The smoke, of course, has a high particulate loading and of itself can be responsible for health distress. Reports of associated illness and increased admissions to medical centres occur at locations well removed from the fires. Chemical reactions also occur in the smoke plume. Many of the reactions and chemicals are the same ones involved in the photochemical smog problem. Smoke plumes from forest fires contain elevated levels of $NO_2$ and VOCs produced from the burning of massive quantities of biomass. Some of the highest levels of ozone and particulate matter have been measured at sites beneath smoke from forest fires when ozone and small particulates are mixed down to ground level. High ozone concentrations associated with smoke plumes from forest fires in Asia have been measured in Alberta, and elevated pollution levels in the southeastern United States have been attributed to forest fires in northern Alberta.

An ongoing concern in Alberta exists with respect to odours. Hydrogen sulphide is the gas of greatest concern, and emissions are regulated in the province. However, industrial accidents that release quantities of $H_2S$ occasionally occur. The accidental blowout of a sour gas well in late 1982 at Lodgepole released large amounts of $H_2S$ that killed two specialists involved in containing the well. Flow rates of sour gas were estimated at four million cubic metres per day. The well flowed uncontrolled for more than 60 days and spread a rotten egg smell over much of central and southern Alberta. The smell was detected more than 1000 kilometres downwind, in Winnipeg, Manitoba. The Lodgepole blowout seemed to galvanize public concerns about air quality in Alberta. Concerns also surround the odours associated with confined feeding operations across the province. Cattle, hog and poultry operations are under continued public scrutiny.

## Air Quality as a Societal Issue in Alberta

Albertans are concerned with maintaining a healthy environment. Governments usually assume the lead in establishing criteria to protect air quality and human health. Alberta has established air quality objectives for many air contaminants, and these are meshed with national goals. The Alberta ambient air quality objectives are expressed in terms of pollutant concentration levels in the atmosphere. Averaging times for pollutant measurements are specified, usually as hourly averages, though daily and annual averages are used. These averaging times are closely linked to impacts— for example, how much time a plant, animal or human will be exposed to the pollutant. The numerical values are established to protect the environment as well as human health.

In addition to the ambient air quality objectives, governments also establish regulations on emissions from various sources of pollutants. For example, the federal government has enacted regulations on emissions from cars and trucks, aimed at reducing the levels of $NO_x$ in cities. At the same time, the provincial government has enacted regulations to reduce emissions from sources such as coal-fired power plants and from flares operated by the oil and gas industry.

In Alberta, the government, industry and the public all have a role to play in dealing with the air quality issue. Several processes are in place to ensure that the three groups of stakeholders undertake joint discussions on many aspects of air quality. Areas for consideration include the setting of air quality objectives, the management of local air quality concerns, and approval of applications for new facilities that may have sources of air pollutants. The groups also evaluate the state of scientific knowledge on many aspects of provincial air quality, and if it is deemed insufficient, they may propose that detailed investigations and research be undertaken.

Alberta Environment operates many air quality monitoring stations across the province.

# Chapter Thirteen:
# The Changing Atmosphere

The atmosphere includes a number of gases that have concentrations at trace levels. Some of the gases have a short residence time in the atmosphere, but some are long lived. Many of these trace gases play an important role because of their physical characteristics or because of their ability to participate in chemical reactions. These gases are important to the earth's environment and, indeed, to the health of the earth's ecosystem. One of the more significant groups of these gases is known as the radiatively active gases, so-called because of the role they play in interacting with electromagnetic radiation—either the radiation emitted by the sun, or by

the earth itself. Some of the gases have a natural origin, and some are only present because people release them into the atmosphere. Ozone and the greenhouse gases play a particularly important role for our planet.

## Stratospheric Ozone

Ozone is a form of oxygen gas that has a dual role in the atmosphere—it is in some ways beneficial and in some ways detrimental. Ozone gas contains three oxygen atoms and is present at low concentrations throughout the atmospheric vertical column from the surface to the stratosphere. Ozone in the stratosphere

## Formation of Stratospheric Ozone

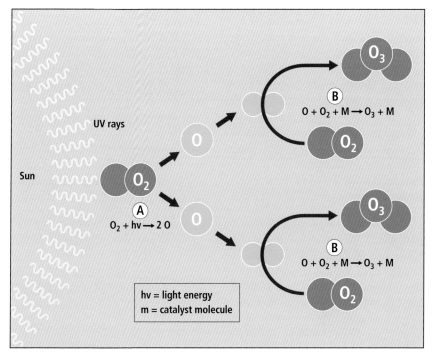

*Fig. 13-1*
A - ultraviolet light strikes oxygen, forming atomic oxygen
B - ozone molecules form in the presence of a catalyst molecule

is produced through natural processes. The ozone molecules absorb much of the short-wave ultraviolet radiation emitted by the sun, protecting life forms from the damaging effects of ultraviolet solar radiation. This property of ozone has been known since the 19th century. Some ozone in the lower atmosphere is produced through the chemical reactions between gases that have natural and anthropogenic origins. This low-level ozone is detrimental to human health and the ecosystem, essentially because ozone is reactive with other substances. Ozone also plays a small role as a greenhouse gas. In the lower atmosphere, ozone contributes to warming of the atmosphere, whereas stratospheric ozone has a cooling effect. The phrase "good up high, bad nearby" describes the dual role of ozone.

### Formation of Stratospheric Ozone

In the upper atmosphere, ultraviolet solar radiation at short wavelengths causes the dissociation of some oxygen molecules into atomic oxygen. The atomic oxygen reacts quickly with oxygen molecules to form the ozone molecule. This reaction also involves a third gas, usually the ever-abundant nitrogen. Nitrogen itself is unaffected, except that it absorbs some of the energy released in the formation of ozone.

The ozone molecule, in turn, absorbs ultraviolet solar radiation and splits apart to form diatomic ($O_2$) and atomic (O) oxygen. This cycle, in which ultraviolet light interacts with oxygen and ozone, is in natural balance in the mid-stratosphere and effectively absorbs harmful short-wave ultraviolet radiation. The intensity of solar radiation is greatest at the equator, and that is where most of the ozone is produced. This ozone is transported toward the polar regions by the large scale, or Hadley, atmospheric circulation pattern (see pp 16–17).

However, substances emitted into the atmosphere are upsetting the natural balance. The gases of concern are the nitrogen oxides, formed in the combustion process, and other gases including the chlorofluorocarbons (CFCs), which contain the chlorine atom, and chemicals containing the bromine atom. Nitrogen oxides ($NO_X$) emitted by combustion in the lower atmosphere normally undergo chemical reactions fairly quickly and do not find their way into the stratosphere, so the concern is with $NO_X$ emitted at high altitudes. Although there were concerns over $NO_X$ emissions from the fleet of supersonic aircraft proposed in the 1970s, there are relatively few aircraft flying at these critical heights, and the resultant ozone depletion from that source has been relatively small to date.

Of much greater concern is the role of CFCs and bromine compounds. When CFCs were first discovered in the early 20th century, they were much valued because they are stable and don't react

A refinery east of Edmonton

A Brewer ozone monitor

with other substances. CFCs were used as propellants in aerosol spray cans, and they replaced the dangerous ammonia gas in most home refrigeration systems. Chemicals containing bromine include methyl bromide, used as an agricultural fumigant, and the halons that are used in fire retardants. The problem with these gases is that they have a long lifespan. When introduced into the lower troposphere, these gases eventually work their way into the upper atmosphere through the action of thunderstorms, which are frequently so vigorous that their tops push into the lower stratosphere. The most energetic thunderstorms are at equatorial latitudes, so most of the penetration into the stratosphere takes place there. These long-lived gases are transported farther into the mid-stratosphere and eventually toward the polar regions by the same Hadley circulation that moves ozone toward the poles.

High energy ultraviolet solar radiation that enters the stratosphere breaks the chemical bonds in the stable compounds, releasing chlorine and bromine atoms into the stratospheric ozone layer. These atoms undergo reactions, forming what are known as catalyst molecules. The catalyst molecules react with the ozone molecules, breaking the ozone apart while the catalyst remains intact. The catalyst goes on to break apart another ozone molecule, and so on. Each chlorine or bromine catalyst molecule can effectively destroy thousands of ozone molecules.

Increased levels of UV radiation can expose us to a greater chance of skin damage from sunburn.

The catalyst reactions that dissociate ozone do not proceed at a uniform rate or equally across the stratosphere. The most pronounced effects are in the polar regions where, during the cold season, stratospheric temperatures can drop to below -80° C. At these temperatures, polar stratospheric clouds containing a mixture of ice, nitric acid and sulphuric acid together with the catalyst chemicals are formed. A reservoir of catalyst chemicals is established in the polar stratospheric cloud layer. The effect is pronounced over the South Pole, where the circulation is effectively concentric around the pole and temperatures can reach the -80° C values. As the earth proceeds in its orbit and polar regions are turned toward the sun, the reservoir warms and the depleting chemicals are released. Destruction of ozone proceeds

rapidly within and adjacent to the reservoir region. The circulation pattern also weakens around the reservoir, allowing ozone-depleted air to mix with ozone-enriched air that is moving poleward from the equator. In the northern hemisphere, with its greater land mass, the circulation pattern has a much greater degree of irregularity than in the southern hemisphere, and temperatures are not as extreme as those over the South Pole. For this reason, there is less ozone destruction in the north polar region than in the south.

The rapid rate of thinning of the ozone layer was first noticed in the south polar region in the mid-1970s. The reduced ozone concentration has become known as the ozone hole, though it is not actually a hole; it is a zone where ozone concentrations are up to 40 percent lower than normal values. These lower concentrations allow much more of the damaging ultraviolet radiation to reach the earth's surface. Because of the increased health risks of solar ultraviolet radiation, governments agreed to take steps to deal with the ozone depletion issue. In 1979, many countries, including Canada, banned the use of CFCs as aerosol propellants. In 1987, an agreement known as the "Montreal Protocol on Substances that Deplete the Ozone Layer" was signed. The agreement required participating countries to reduce their production of CFCs. The agreement was later strengthened to phase out all CFC production.

Actions taken under the Montreal protocol are having a positive effect. Measurements in the high atmosphere now show that the concentrations of some of the ozone-depleting substances are slowly starting to decrease. There is, however, a long way to go. It will take until the middle of the 21st century for chlorine levels to return to the values measured in the 1970s when the south polar ozone hole first started appearing. Current projections by Canadian scientists indicate that global total ozone levels will return to 1960s averages by about 2060. The recovery assumes that countries will continue to adhere to the terms of the Montreal protocol and that no other long-lived substances will be found to cause ozone depletion. The possibility of increased $NO_X$ emissions is still a concern because consideration is being given to allowing more high-flying supersonic transport aircraft. Another concern is the potential effect that global warming may have on stratospheric ozone in the polar regions. This is an ongoing topic of Canadian research.

## Stratospheric Ozone and Alberta

In Alberta, the thinning ozone layer will continue to be a concern. In late winter to early summer, when ozone thinning is most noticeable over Alberta, the public is vulnerable to increased levels of ultraviolet radiation. At that time of year, we need to take extra protective measures. This is especially important for those people who enjoy spring skiing. The intensity of UV radiation at alpine levels is enhanced by the thinner atmosphere and by the additional solar reflection off snow surfaces.

## Climate Change

Albertans are well aware of the changes in weather from day to day. We know what to expect from our climate from year to year, but we also know that there will be some degree of variability. Some winters have more snow, some summers are hotter and some have more rain than others. Some years seem to have more severe summer storms, whereas some years are quite placid. This wide range of variability seems to be the norm. When averages of temperature, rainfall or any of the other atmospheric parameters are calculated at an observing station or are averaged over a region, much of the variability disappears. However, what is apparent from these averages is that there is an upward trend in temperature with time.

People are finding ways to protect themselves from harmful sun overexposure.

Is the climatic average moving to some new state, and if so, why? What will this new state mean for our lifestyle, for the environment? Can we change the trend? What measures will society need to take to deal with the change? What will be the cost of these measures? These are the many questions that are raised by the climate change issue.

Temperature is the most useful parameter to use in studies of climate and its changes. Temperatures are reasonably consistent over an area, and the temperature record can usually be readily established. The instrumented record only goes back about 120 years in Alberta. It is somewhat longer at other locations across Canada (back to about 1840 in Toronto) and only marginally longer at some sites around the world. Prior to the instrumented record, what can we say about the temperature and climatic conditions? Scientists have developed a number of techniques to deduce past conditions. We have anecdotal information about storms, cropping practices and settlement patterns that allow us to infer what temperature conditions were like over the last few centuries. From the examination of tree ring widths, we have reasonably direct evidence of the length and productivity of growing seasons as far back as several hundred years. The

analysis of lake and ocean bed deposits provides a wealth of information about plant forms that were present on adjacent lands or of plants and animals that inhabited the water bodies. This evidence from around the world helps push the climatic record back several thousand years. Carbon dating is a sophisticated tool that is used to look back about 40,000 years, and it is accurate to within about 100 years. The characteristics of ice in glaciers can also be used to infer what temperatures were like at the time the snow fell on glacier surfaces. Ice cores extracted from Antarctica and from Greenland glaciers have been used to determine temperatures back over 600,000 years.

The climate has undergone many natural transitions in the past. Alberta was once covered with lush tropical growth, which is why we now have such a wealth of coal, gas and oil reserves. Ice sheets have spread across the North American continent with a fair degree of regularity, every 100,000 years or so. There are several theories that describe why past climatic changes occurred. The first theory relates to the all-important link between the earth and the sun. The earth travels around the sun in a slightly elliptical orbit, and the earth's axis of rotation is tilted at an angle of 23 degrees. The Serbian mathematician

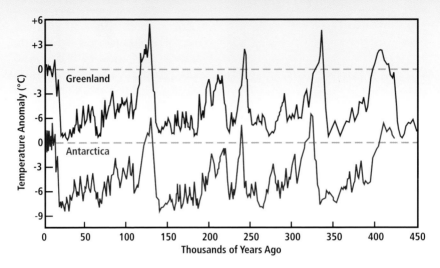

Fig. 13-2

Milutin Milankovitch examined the regular changes in the degree of orbit elipticity, rates of change in the axis tilt and changes in the time of year that the earth is closest to the sun. With this information, he developed a schedule that could explain the onset and termination of past ice ages.

Shorter-term fluctuations in solar energy output are well known. The sun goes through a cycle in which sunspots come and go every 11 years. When the number of sunspots is at a maximum, there is a slight elevation in the rate of solar energy output. The increase in radiation is only slightly above the normal levels, though, amounting to less than 0.2 percent over the 11-year sunspot cycle. The sun's energy output may have deviated for periods of a half-century or so when there was an almost total absence of sunspots. The associated solar output decrease has been linked with a cold period in Europe in the 16th and 17th centuries, known as the Little Ice Age.

## Earth's Energy Balance and the Greenhouse Effect

The temperature of the atmosphere is linked to the balance between incoming solar radiation and the properties of the earth's surface and the surrounding atmospheric envelope. Structures, water and life forms on the earth's surface absorb a large fraction of the total incoming solar radiation. Objects that absorb radiation are warmed, and they in turn radiate energy in the infrared wavelength. Much of this infrared radiation is directly absorbed by radiatively active gases, which then warm and in turn radiate energy. All this energy transfer, absorption and retransmission are in a general state of balance over the earth.

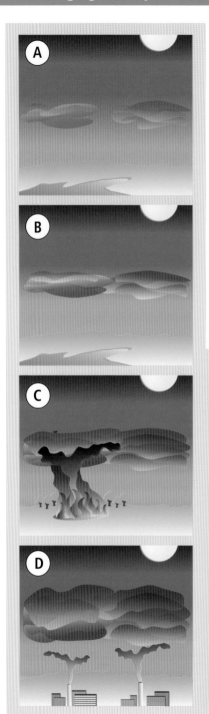

This balance is maintained by the meridional circulation. The global energy balance results in an annual average temperature of about 16° C over the entire planet. Of key importance is the role of the radiatively active gases. In 1824, famous French physicist and mathematician Joseph Fourier first proposed the concept that gases in the atmosphere trap heat from the sun. This process has been termed the greenhouse effect, though that is something of a misnomer—greenhouses prevent the mechanical movement of heated air, whereas atmospheric gases reduce the radiative loss of heat. It is estimated that without greenhouse gases, the average global temperature would be about -18° C, which is too cold to support life as we know it. Mars has a thin atmosphere with a low concentration of water vapour and

## The Greenhouse Effect
*Fig. 13-3*

A - sunlight travels though the atmosphere and strikes the earth's surface, which warms and radiates infared energy.

B - most of the infared heat is absorbed by the so-called greenhouse gases in the atmosphere. The gases warm and then radiate heat energy back to the earth.

C - burning forests and other fires release greenhouse gases, especially carbon dioxide, into the atmosphere. This is the natural greenhouse effect, which has given the earth an average surface temperature near 16° C.

D - additional carbon dioxide and other greenhouse gases are released to the atmosphere from fossil fuel combustion in factories and power plants and from transportation. This anthropogenic greenhouse effect is warming the earth and atmosphere at unprecedented rates.

Burning of fossil fuels, such as coal, releases carbon dioxide into the atmosphere.

other greenhouse gases; temperatures on the planet are below -50°C. Venus, on the other hand, has a dense atmosphere with high proportions of radiatively active gases. The greenhouse effect on Venus maintains temperatures at above 400°C.

Many gases play a role in the natural greenhouse effect on earth, and the most important is water vapour. Water molecules are continuously cycling between the atmosphere and water surfaces, changing phase from vapour to liquid to solid. The evaporation, sublimation and condensation processes all involve substantial energy exchanges, which drive much of the weather we experience. The average water content of the atmosphere changes little and is kept in balance through the evaporation-precipitation cycle. However, how much water the atmosphere can hold is dependent on temperature—a warmer atmosphere tends to hold more water vapour, providing a positive feedback mechanism to heat the atmosphere further.

Other important greenhouse gases are carbon dioxide ($CO_2$), methane ($CH_4$) and nitrous oxide ($N_2O$). These gases have a natural origin, but they are also produced through human activity. Carbon dioxide is released from combustion processes such as forest fires and through fossil fuel combustion in heating systems, engines and boilers. Carbon dioxide is

215

also produced in respiration and through the aerobic decay of vegetation. Methane is produced through anaerobic decay of vegetation and in the digestive tract of ruminants. It is the primary constituent of natural gas, it is contained in coal beds and it is a minor by-product of the combustion of carbon-based fuels. Nitrous oxide is primarily a product of combustion, though it is also released to the atmosphere when nitrogen-based fertilizers break down. All of these gases have atmospheric lifespans that are much longer than that of water vapour, and this is what makes them of particular concern as greenhouse gases because they tend to accumulate in the atmosphere.

There are other gases that have no natural form in nature but are only produced by humans. Substances such as CFCs, halons and several others are in this category and are potent greenhouse gases, in part because of their long atmospheric lifespans.

# The Trend in Global Greenhouse Gases

Atmospheric $CO_2$ concentrations have been measured for less than a century. Measurements taken at Mauna Loa, Hawaii, since 1956 show a steady upward trend, with mean annual atmospheric concentrations increasing from about 300 parts per million (ppm) to the present value of 380 ppm. Similar trends are noted at Canadian sites and at other locations around the world. Analysis of the atmospheric gases trapped in glaciers and ice sheets provides a longer record. Ice cores from the south polar ice sheet have made it possible to reconstruct the atmosphere's $CO_2$ concentration extending back more than 600,000 years. Data from these cores shows that $CO_2$ concentrations were about 270 ppm prior to industrialization. During periods of major global glaciations, global $CO_2$ concentration levels were about 200 ppm.

Glaciers are retreating in a warming climate.

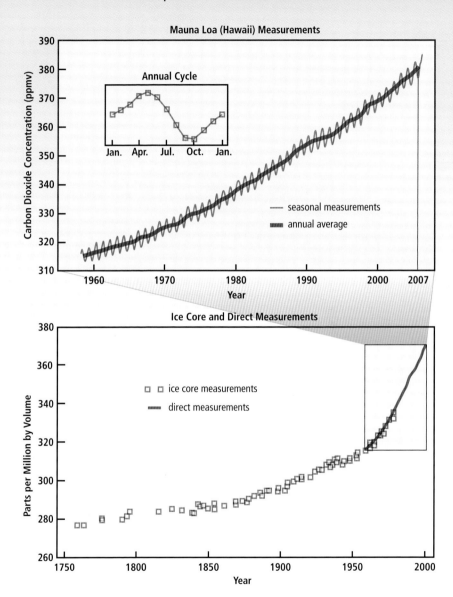

Fig. 13-4 Atmospheric carbon dioxide concentrations as determined from ice cores show a steady but modest increase from 1750 to 1950. Since 1960, instrument measurements and ice core data show the rate of increase has accelerated. Global values are now over 380 ppm.

## Global Temperatures

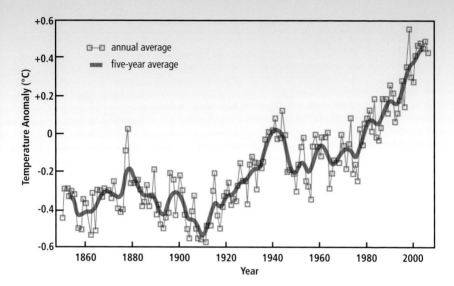

Fig. 13-5 The squares are average temperatures computed from thousands of surface stations around the globe. The data reflect averages for the period from 1961 to 1990. The smooth curve is the five-year running mean of the plotted data.

During the warm interglacial periods, including the present one, $CO_2$ levels attained peak values of 270 to 300 ppm. Methane and nitrous oxide concentrations show similar long-term trends. In summary, the present concentrations of all the key greenhouse gases are far higher than levels over the past 600,000 years.

## Concerns about Warming

In 1896, while investigating the causes of periodic ice ages, Swedish chemist Svante Arrhenius quantified the relationship between the average concentrations of greenhouse gases and average global temperatures. He concluded that a doubling of atmospheric carbon dioxide concentrations would likely result in an average global warming of 5° to 6° C.

Arrhenius realized that the carbon dioxide introduced by industrial emissions was in fact accumulating in the atmosphere, but he concluded that, given the emission rates at the turn of the 20th century, a doubling of $CO_2$ concentrations would be unlikely for 3000 years. In reality, emissions from new sources of industrial emissions are being produced at much greater rates than he envisioned, and a doubling might be expected to occur within 100 years.

There is clear evidence that the earth has undergone a steady warming over the past century. The scientific community has been working to confirm the climate trend and to determine the reasons. Careful examination and re-examination of temperature data collected around the globe reinforces that a warming is underway. The temperature record shows that since the end of the 19th century, the average global temperature has risen about 0.6° C. This warming has not been steady. The instrument record shows that a period of warming from about 1910 to 1940 was followed by a short period of cooling until the mid-1970s. The warming since then has been steady. The variability of the trend has caused much controversy. Some brief cooling periods can be explained by volcanic eruptions such as Mount Pinatubo in 1991, but not all the variability has been explained. Nevertheless, the general trend is decidedly upward, and most scientists agree that this is because of excessive greenhouse gas emissions.

Data has been compiled for the northern hemisphere, consisting of an amalgam of measured temperature records and temperatures inferred from tree ring growth, coral growth and ice core analyses. The period from about 1000 AD to 1900 AD showed a slight cooling. Over the past century, however, the warming has been remarkable.

## Changes in Alberta

In Alberta, the temperature record extends back to the late 19th century. When annual temperatures are plotted, there is considerable variation from year to year, but the temperature data shows evidence of a warming trend. The temperature record for sites in southern, central and northern Alberta show

## Reconstructed Temperatures

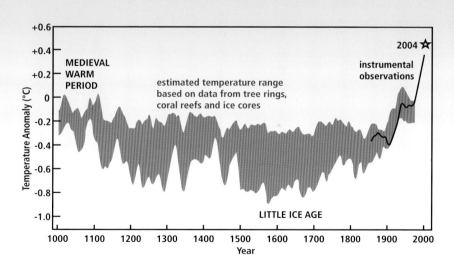

*Fig. 13-6* The shaded region shows the range of uncertainty in temperatures inferred from tree rings, coral reefs and ice cores. The solid line is the instrumented record. All data reflect the global averages from 1961 to 1990.

# Annual Temperatures at Alberta Locations

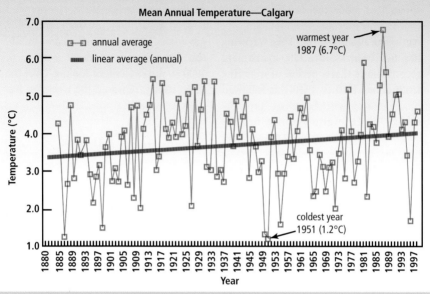

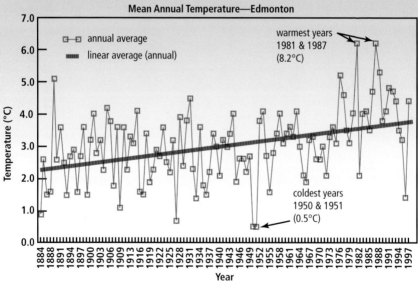

*Fig. 13-7* The squares are the annual average temperatures and show a large year-to-year variability. The solid blue line is a single linear best-fit curve for the full data period at each location.

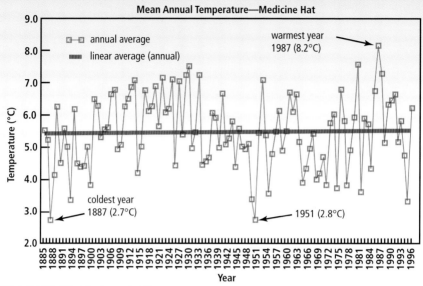

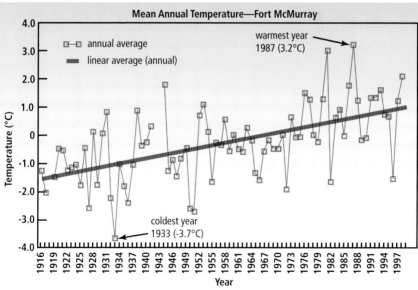

221

elements of the same trends noticed elsewhere in the northern hemisphere. As is shown in Figure 13-7, there has been an overall upward trend in the annual temperature for four stations in Alberta, and the north appears to be warming at a greater rate than the south. Calgary has warmed by about 0.6° C and Edmonton by 1.3° C throughout the 20th century. Fort McMurray has warmed by 3.5° C over a somewhat shorter period. Medicine Hat, on the other hand, shows only a slight warming trend of 0.1° C throughout the 20th century. If the climate record is examined for shorter periods, the Alberta stations show a general upward temperature trend until the 1930s, followed by cooling to the early 1970s, then a resumption of warming.

> **Love comforteth like sunshine after rain.**
>
> —Shakespeare,
> *Venus and Adonis*

The precipitation data shows a mixed trend over the 20th century across Alberta because there is a large range of variability in precipitation. One way to make some sense from the data is to examine what is known as the climate normals. These are the averages of climatic weather observations for a 30-year period at a specific location. When normals from one period are compared to those from another, locations in southern Alberta show a trend toward drier conditions over the last half of the 20th century. However, in central and northern Alberta, there is little evidence of any meaningful trend in precipitation.

# Future Climate

Predicting future climate is a challenging task. The climate system is composed of several components, including the atmosphere, the hydrosphere (oceans, lakes, wetlands and rivers), the earth's cryosphere (glaciers, floating ice) and the biosphere (including all living things). These components of the climate system interact and influence each other. Components of the climate system that have a physical science basis can be described by mathematical relationships. A representation of the climate system can then be made by means of a complex mathematical model formulated to operate on computers. This is known as a climate simulation model. These equations depend on time, so the model can be used to examine how the climate system changes with time. Scientists use this model to develop scenarios of the future climate.

Climate simulation models work much like the weather prediction models described in an earlier chapter, but there are several important differences. Weather prediction and climate simulation models both use the laws of fluid physics to describe the atmosphere and its changes with time. Climate models, however, take into account many additional processes that are not important for short-range weather forecasting. Weather prediction models start from an initial state of the atmosphere and run forward to a few days in the future. Climate models run forward for decades, even out to 100 years or more. The climate models run on a coarser scale than weather prediction models—usually with a horizontal grid spacing of hundreds

rather than tens of kilometres—and therefore the climate detail is relatively coarse. All the complexities of the linkages between the atmosphere, the hydrosphere and the cryosphere must be calculated using relationships appropriate to the grid scale of the model. The number of calculations necessary to undertake a climate simulation can only be accomplished in a reasonable time frame by using the most powerful computers available. Only a few centres in the world have sufficient computing power and scientific expertise to undertake numerical climate simulations. Some of the more prominent centres include the Hadley Centre of the British Meteorological Office, the Geophysical Fluid Dynamics Laboratory (GFDL) of the National Oceanographic and Atmospheric Administration of the United States, the National Center for Atmospheric Research (NCAR) located in Colorado, and the Earth Simulator Center in Japan. In Canada, Environment Canada conducts climate simulation modelling with a group located at the Canadian Centre for Climate Modelling and Analysis in Victoria, BC.

The output of the climate simulation models can be displayed either in pattern form (i.e., maps) or by displaying the value of some climatic parameter that is computed by the model. Average surface temperatures over the globe, over a hemisphere or even over a region are often displayed. For meteorologists to have confidence that the models are correct, the models must accurately depict our present climate. A good test of their validity is to see how they perform in simulating past climate trends.

## Model Simulations of Past Climate Trends

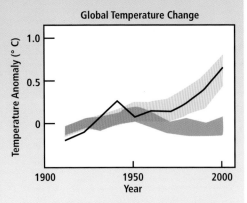

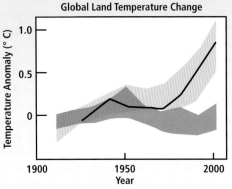

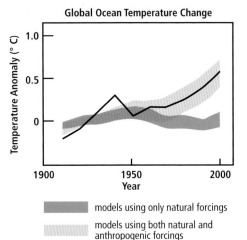

models using only natural forcings

models using both natural and anthropogenic forcings

observations

Fig. 13-8

Figure 13-8 shows how 14 models have performed in simulations of worldwide surface temperature over the oceans and land services throughout the past century. Also shown is the average of observations calculated in the same areas over the same period. The ranges of the model simulations, shown as shaded bands, depend on the assumptions used in the formulation of the model physics. The graph shows that models best replicate the observed global temperature trend when increases in anthropogenic greenhouse gas emissions are incorporated in model simulations.

In 1988, the United Nations established the Intergovernmental Panel on Climate Change (IPCC) to investigate whether climate change is being caused by human activity. The emphasis over the past three decades has been on the role of greenhouse gases in changing the short-term climate. Climatic simulation models are used to undertake experiments using a range of greenhouse gas emission scenarios. Figure 13-9 shows the results of several emission scenario experiments undertaken using a number of climatic simulation models. The models show that there is a wide range of possible outcomes for the future climate, depending on the emissions scenario. The models project that by the end of the 21st century, the earth will have warmed between 1.1° C and 6.4° C relative to the global temperature at the end of the 20th century.

## Alberta's Future Climate

The present trends in climatic observations together with climate model simulations suggest that the world's climate will undergo further warming during this century. A few general observations can be made about weather trends in Alberta based on climate projections from the Canadian climate simulation model. The province will likely undergo a continued gradual warming throughout this century with a greater warming rate in the north than in the south. One can anticipate more frequent warm spells in summer. In winter, a trend toward less frequent cold spells is expected. This does not mean that we will not experience any more cold days or even cold winters.

Occasional cold winters will still happen, but they can be expected to occur less frequently than we have seen in the past.

The model simulations suggest little or no change in average precipitation amounts, but there may well be a trend to more frequent and extreme precipitation events. The combination of little change in annual precipitation together with general warming suggests a greater frequency of extreme summer heat waves and a trend toward a more arid climate.

Some speculative studies have been undertaken to look at the impacts of an increasingly arid climate across western Canada. There is the potential for the winter snowfall season to be shortened,

## Multi-modal Averages and Assessed Ranges for Surface Warming

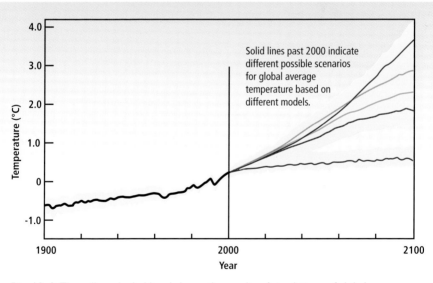

Solid lines past 2000 indicate different possible scenarios for global average temperature based on different models.

*Fig. 13-9* The yellow shaded band shows the results of simulations of global average surface temperature from many different climate simulation models. The solid curve prior to 2000 is the observed temperature. Each solid line past 2000 is the average of model simulations for one emission scenario.

A thinning ozone layer poses a threat of sun damage to unprotected skin.

with a decreased snowpack and earlier melt. This would place additional stress on the hydrological system in Alberta. Alberta's boreal forests may be subjected to additional stress, and climate conditions may favour the northward progression of grasslands. The increasing aridity and potential for more intense and frequent summer storms may result in increased frequency of summer forest fires. As well, there is potential for enhanced pest infestation in Alberta's forests because of more moderate winter temperatures. Changes in recreational resources could also be anticipated with a more arid climate. Prairie wetlands might continue to decline, which would further stress migratory bird populations. The potential loss of mountain snowpack may necessitate a greater reliance on artificial snowmaking to maintain Alberta's important recreational skiing and snowboarding industry.

Because climate models typically use a fairly coarse grid, simulating climate changes at a regional scale is a major challenge. Regional climate models are being developed to deal with this issue and are being tested throughout western Canada. The outputs from the Canadian model are eagerly anticipated. More definitive studies will undoubtedly follow, providing better resolution of climate trends and impacts. This will then allow for an improved understanding of the options available and the costs associated with accommodating a changing climate.

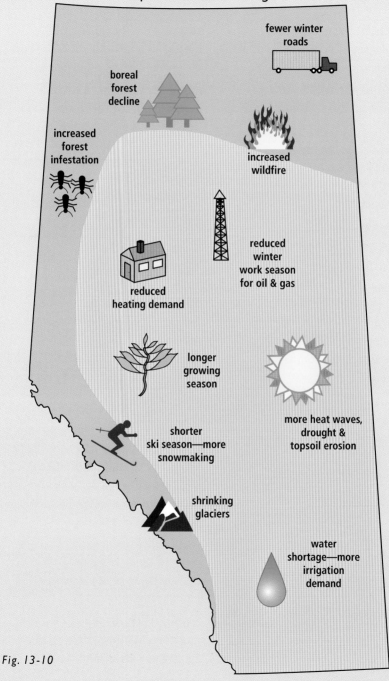

Impact of Climate Change

fewer winter roads

boreal forest decline

increased forest infestation

increased wildfire

reduced winter work season for oil & gas

reduced heating demand

longer growing season

more heat waves, drought & topsoil erosion

shorter ski season—more snowmaking

shrinking glaciers

water shortage—more irrigation demand

*Fig. 13-10*

# Glossary

**accretion:** the growth of a hydrometeor through collision with supercooled cloud drops.

**aerology:** the study of the atmosphere throughout its vertical extent.

**air mass:** an extensive body of air within which temperature and moisture at a constant height are essentially uniform.

**ambient air quality:** air pollutant concentration levels for outdoor air.

**anticyclone:** an atmospheric pressure system that has relatively high pressure at its centre and, in the northern hemisphere, has winds blowing around the centre in a clockwise direction; also called, simply, a high.

**azimuth angle:** the angle in degrees from north measured in a clockwise direction.

**blue jet:** a weakly luminous blue-coloured discharge that moves upward from the top of a thunderstorm.

**cirrus:** a principal cloud type in the form of thin or feather-like clouds in the upper atmosphere, made up of ice crystals.

**cumulus:** a cloud in the form of detached domes or towers that appear dense and have well-defined edges; has a flat, nearly horizontal base with a bulging upper portion that resembles cauliflower.

**cup anemometer:** a wind-measuring instrument consisting of three or four hemispherical or conical-shaped cups mounted on horizontal arms that rotate around a vertical axis.

**cyclone:** an atmospheric pressure system that has relatively low pressure at its centre and, in the northern hemisphere, has winds blowing around the centre in a counterclockwise direction; also called, simply, a low.

**discretization:** the method of representing a continuously varying parameter with a discrete set of values.

**front:** the boundary region along the earth's surface that separates air masses; frontal types are characterized by the nature of the cold air mass, the direction of the movement of the front and its stage of development.

**frontal lift:** the upward displacement of an air mass of lower density by an air mass with a higher density.

**frontal surface:** the interface above the earth's surface that separates different air masses.

**fulgurite:** a glassy tube formed when a lightning stroke terminates in dry, sandy soil and fuses the sand.

**geostrophic wind:** the air motion that results from air pressure differences.

**gradient:** the change of a meteorological parameter per unit of distance.

**graupel:** a form of ice created in the atmosphere when supercooled water droplets coat an ice crystal.

**hydrometeor:** a general term for any type of water or ice particle.

**isobar:** a line on a chart or diagram drawn through points that have the same barometric pressure.

**jet stream:** a zone of relatively strong winds concentrated within a narrow band in the atmosphere.

**lapse rate:** the rate of decrease of atmospheric temperature with height.

**mares' tail:** a cirrus cloud that has long, wispy strands extending from a tufted end.

**mesocyclone:** an updraft column of air with a diameter of 2 to 10 kilometres that rotates around a vertical axis in a large thunderstorm; the rotating column can lead to the formation of tornados.

**mesopause:** the atmospheric layer located at a height of around 90 kilometres, at the top of the mesosphere, where the coldest atmospheric temperatures occur.

**mesosphere:** the atmospheric layer above the stratosphere, characterized by temperatures that decrease from the base of the layer to the top.

**N-region:** the negatively charged region in the lower portion of a thunderstorm cloud

**planetary (long) waves:** a wave-shaped pattern in the westerly wind belt characterized by long length; four or five waves are typically found around the hemisphere, and the wave pattern generally moves slowly toward the east with a speed much slower than the speed of the westerly winds.

**planetary boundary layer:** the bottom layer of the troposphere in contact with the earth's surface.

**plow winds:** strong, straight line winds associated with the outflow of a strong thunderstorm.

**p-region:** the positively charged region near the bottom of a thunderstorm cloud.

**P-region:** the positively charged region near the top of a thunderstorm cloud.

**pressure gradient force:** the force acting on air caused by differences in air pressure; the strength of the force is proportional to the difference in air pressure.

**psychrometer:** an instrument that measures humidity using two thermometers, one of which has its sensing bulb covered by a water-saturated muslin jacket.

**retrogression:** westward movement of weather systems at middle latitudes, instead of the usual eastward movement.

**red sprite:** a large-scale luminous flash that appears directly above an active thunderstorm and extends from a cloud top to heights of about 90 kilometres; it is delayed a short time after a cloud-to-ground lightning stroke.

**ridge:** an elongated area of high atmospheric pressure on a weather chart.

**rime:** milky colored ice formed when super-cooled water drops freeze on contact with a surface and create a mass of tiny balls with air spaces between them.

**smog:** originally defined as a mixture of smoke and natural fog, but now describes a mixture of air pollutants that are emitted by automobiles and industrial sources and are acted upon by the sun.

**snow pillow:** an instrument used to estimate snow pack by measuring the pressure exerted by a mass of overlying snow.

**solar wind:** a stream of charged particles flowing outward from the sun.

**stepped leader:** the column of highly ionized air that intermittently advances downward from a thunderstorm cloud and establishes a channel for a lightning stroke.

**stratopause:** the top of the stratosphere, characterized by a reversal of the rate of change of temperature with height.

**stratosphere:** the atmospheric layer from about 10 to 50 kilometers above the earth, characterized by temperatures that warm from the base to the top of the layer.

**stratus:** a cloud layer with a fairly uniform base.

**streamer:** an upward-advancing column of highly ionized air that moves from a point on the earth's surface toward a stepped leader.

**suction vortex:** a small-scale secondary vortex within a tornado core that moves around a central vertical axis.

**supercell:** a large, severe thunderstorm with rotating updraft that usually lasts several hours and produces heavy rain and hail, strong winds and occasionally spawns tornados.

**synoptic scale storm:** a large-scale atmospheric system or weather event that has a horizontal dimension of greater than 300 kilometres.

**thermistor:** a temperature sensor in which the electrical resistance varies in proportion to the temperature; the name is a combination of "thermal" and "resistor."

**thermosphere:** the upper atmosphere above 100 kilometers, characterized by low gas density and temperatures increasing with height.

**tripole distribution:** the characteristic distribution of positive and negative charge centers in thunderstorm clouds.

**tropopause:** the transition zone in the upper atmosphere between the troposphere and the stratosphere, characterized by an abrupt reversal of the thermal lapse rate.

**troposphere:** the atmospheric layer from the earth's surface to about 8 to 12 kilometers, characterized by decreasing temperature with height and substantial water vapor; the region in which most weather occurs.

**trough:** an elongated area of low atmospheric pressure on a weather chart.

# Resources

Angle, R.P. and S.K. Sakiyama. (1991). *Plume Dispersion in Alberta*. Edmonton: Standards and Approvals Division, Alberta Environment.

Barry, R.G. and R.J. Chorley. (1971). *Atmosphere, Weather and Climate*. London: Butler & Tanner Ltd.

Bates, David V. and Robert B. Caton. (2002). *A Citizen's Guide to Air Pollution* (2nd ed.). Vancouver: David Suzuki Foundation.

Bullas, John M. (1990). The Climate of the Prairie Provinces 1961–1990: A Review. Unpublished document. Environment Canada.

Bullas, John M. and Alton F. Wallace. (1988). The Edmonton Tornado, July 31, 1987. Preprints, 15th Conference on Severe Local Storms, Baltimore. Boston: American Meteorological Society.

Cessna Pilot Center. (1984). *Canadian Manual of Flight*. Denver: Jeppesen & Co.

Chen, Thierry. (1988). Dust Storm of May 23, 1988. Western Region Technical Notes 88-N-082. Unpublished document. Environment Canada.

Chetner, S. and Agroclimatic Atlas Working Group. (2003). *Agroclimatic Atlas of Alberta 1971–2000* (Agdex 071-1). Edmonton: Alberta Agriculture, Food and Rural Development.

Drew, John. (1855). *Practical Meteorology*. London: John Van Voorst.

Fletcher, N.H. (1962). *The Physics of Rainclouds*. New York: Cambridge University Press.

Flohn, Hermann. (1969). *Climate and Weather*. New York: McGraw-Hill Book Company.

*From the Ground Up*. (1987). Ottawa: Aviation Publishers Co. Ltd.

Gdezelman, Stanley David. (1980). *The Science and Wonder of the Atmosphere*. New York: John Wiley & Sons.

Hage, K. (2003). On destructive Canadian Prairie windstorms and severe winters. *Natural Hazards*, 29, 207–228.

Hare, F. Kenneth and Morley K. Thomas. (1974). *Climate Canada*. Toronto: John Wiley & Sons Canada Ltd.

Harrison, Louis P. (1942). *Meteorology*. New York: National Aeronautics Council, Inc.

Holton, James. R, Judith A. Cury and John A. Pyle (Eds.). (2002). *Encyclopedia of Atmospheric Sciences*. San Diego: Academic Press.

Intergovernmental Panel on Climate Change. (2007). Summary for Policymakers. In *Climate Change 2007: The Physical Science Basis* (contribution of Working Group I to the Fourth Assessment Report of the Intergovernmental Panel on Climate Change). Cambridge: Cambridge University Press.

Klivokiotis, P. and R.B. Thomson. (1986). *The Climate of Calgary.* Ottawa: Environment Canada, Supply and Services Canada.

Stanley, George F.G. (1955). *John Henry Lefroy: In Search of the Magnetic North.* Toronto: MacMillan Company of Canada Ltd.

Longley, R. (1972). *The Climate of the Prairie Provinces.* Ottawa: Environment Canada.

Meteorological Branch. (1964). *Weather Ways.* Ottawa: Department of Transport, Queens Printer and Controller of Stationery.

Neault, Gerard and Arjen Verkaik. (1983). The Rocky Mountain House Tornado. Western Region Technical Notes 83-N-034. Unpublished document. Environment Canada.

Olson, Rod. (1985). *The Climate of Edmonton.* Ottawa: Environment Canada, Supply and Services Canada.

Paruk, B.J. (1988). Spring Blizzard 1988. Western Region Report. Unpublished document. Environment Canada.

Phillips, David. (1990). *The Climates of Canada.* Ottawa: Supply and Services Canada.

Rakov, Vladimir A. and Martin A. Uman. (2003). *Lightning: Physics and Effects.* Cambridge: Cambridge University Press.

Shaefer, Vincent J. and John A. Day. (1981). *A Field Guide to the Atmosphere.* Boston: Houghton Mifflin Co.

Spiegel, Herbert J. and Arnold Gruber. (1983). *From Weather Vanes to Satellites.* New York: John Wiley & Sons.

Strong, G.S., B. Kochtubajda, P.W. Summers, J.H. Renick, T.W. Krauss, R.G. Humphries and E.P. Lozowski. (2007). 50th anniversary of hail studies in Alberta: accomplishments and legacy (Bull.). *Canadian Meteorological and Oceanographic Society, 35,* 3–28.

Uman, Martin A. (1986). *All About Lightning.* New York: Dover Publications, Inc.

Wallace, A.F. (1987). Track of the Edmonton Tornado, July 31, 1987. Western Region Technical Notes 87-079. Preliminary report. Environment Canada.

Weisberg, Joseph S. (1981). *Meteorology: the Earth and Its Weather.* Boston: Houghton Mifflin Co.

Williams, Jack. (1992). *The Weather Book.* New York: Vintage Books.

Williamson, S.J. (1973). *Fundamentals of Air Pollution.* Reading: Addison Wesley Publishing Co.

Zielinski, Gregory A. and Barry D. Keim. (2003). *New England Weather, New England Climate.* Lebanon: University Press of New England.

# Index

# Photo Credits

BC Parks / Gail Ross 206; Digital Vision Ltd. 38, 39, 216; Kristina Friesen 5, 29; Bill Hume 36, 48, 49, 50, 51a, 163b, 175, 184, 185, 186a, 186b, 209, 240; iStockphoto.com / Hans Caluwaerts 210; iStockphoto.com / Sebastien Cole 138; iStockphoto.com / Glenn Frank 8, 152, 180; iStockphoto.com / Constance McGuire 187; iStockphoto.com / Slavoljub Pantelic 212; iStockphoto.com / Andrew Penner 142; iStockphoto.com / Jason Verschoor 101; iStockphoto.com / Shane White 211; iStockphoto.com / Serdar Yagci 181; Jupiterimages Corporation 9, 10, 30, 31, 32, 34, 40, 45, 51b, 65, 98, 104, 137, 141, 183, 194, 195, 196, 199, 215, 223, 226, 230; Randy Kennedy 182, 198, 208; Alister Ling 33, 44, 47a, 52, 72, 78a, 82, 99, 108, 110; Edward P. Lozowski 64, 150; NASA / Veres Viktor 47b; NOAA Photo Library 60, 66, 92, 95; NOAA Photo Library / Grant W. Goodge 41; NOAA Photo Library / Mr. Sean Linehan 162; PhotoDisc, Inc. 75; Dennis Quintilio 136; Jim Renick 188, 189, 193; Heather Rombough 35; Ian Sheldon 1, 4, 25, 107, 190; The Edmonton Journal / Steve Simon 146; TransAlta Energy Corporation 83; Weather Modification Inc. 191; Gary Whyte (www.garywhyte.com) 46, 54, 114, 134, 139, 197; Brian Wiens 204a, 204b

© Her Majesty The Queen in Right of Canada, Environment Canada, 2008. Reproduced with the permission of the Minister of Public Works and Government Services Canada: 77, 155, 157, 158, 159a, 159b, 160a, 160b, 161, 163a, 165, 166, 167, 168, 169a, 169b, 179, 203, 205;

© Her Majesty The Queen in Right of Canada, Environment Canada, 2008. Reproduced with the permission of the Minister of Public Works and Government Services Canada / NOAA: 102, 172, 173;

Creative Commons:
changvoli 153; Bryan Partington 84; SqueakyMarmot 111

Creative Commons, Attribution Share Alike:
Dan Huntington 43; jurvetson 97; Luna04 78b; United States Air Force / Senior Airman Joshua Strang 80

Public Domain 76

Bill Hume started his career in atmospheric sciences in 1971 with Environment Canada in Toronto. He attained a Master's degree in meteorology from the University of Alberta in 1975 and has spent most of his working career in Edmonton and Calgary. He has experience in most aspects of meteorology including weather forecasting, climatology, air quality science and atmospheric research. He has taught meteorology at the University of Alberta and served on several national committees including the Canadian Meteorological and Oceanographic Society. Bill and his wife Judy have two married children and reside in Edmonton.